ULTRAVIOLET LIGHT AND FLUORESCENT MINERALS

ULTRAVIOLET LIGHT AND FLUORESCENT MINERALS

Understanding, Collecting and Displaying Fluorescent Minerals

Thomas S. Warren
Sterling Gleason
Richard C. Bostwick
Earl R. Verbeek

Thomas S. Warren
Publisher

ISBN No. 0-9635098-0-2

Cover Photo

The zinc mines of the Franklin-Sterling Hill area in northern New Jersey have long been famous for the variety and brilliance of their fluorescent minerals. Shown on the front cover is typically spectacular zinc ore from Sterling Hill, with veins of green-fluorescent willemite cutting red-fluorescent calcite. The abundant black, nonfluorescent grains are franklinite. This pair of slabs, each 8 x 5 inches in size, was sawn from a single specimen. The slabs were photographed under shortwave ultraviolet light (top) and in daylight (bottom) by David Grigsby. They are now in the collection of the Reverend Dr. William E. Foley, who has kindly given permission to use the photos.

This book may be ordered from the distributor:
Williams Minerals
R.R. 1, Box 77
Dutch Hollow Road,
Rio, West Virginia 26755
Telephone: (304) 897-6003

DEDICATION

This book is dedicated to my two daughters, Bethany Griffiths and Virginia Hiramatsu, whose devotion and cooperation have made life more meaningful and worthwhile.

THOMAS S. WARREN

THE INFORMATION IN THE FOUR CHAPTERS OF THIS BOOK

Chapter I: The Magic Of Ultraviolet Light

Thomas S. Warren

This chapter explores the mysterious properties of ultraviolet light: its unique effects on human health; its many uses in medicine, criminology, and agriculture; and its almost magical property of causing fluorescence in a wide range of minerals. The author, founder and long-time president of Ultra-Violet Products, Inc., has been associated with the fluorescent-mineral hobby for almost 60 years. He explains the different types of ultraviolet lamps, their relation to the familiar fluorescent lamps used for household and office lighting, and how recent advances in manufacturing have greatly expanded our enjoyment of mineral fluorescence. Hints on selecting appropriate lamps for field collecting, study, and display introduce the reader to the practical side of the hobby.

Chapter II: The Fascination of Fluorescent Minerals

Sterling Gleason

This chapter, a collection of excerpts from Gleason's original book *Ultraviolet Guide to Minerals,* introduces hobbyists to the world of fluorescent minerals through imaginary prospecting trips. Along the way the reader learns that it is not only minerals that fluoresce, but also common household items, plants, clothes, fossils, and even the human body. In addition, the chapter touches on the use of portable ultraviolet lamps in prospecting for ores of tungsten, uranium, lead, and zinc, and it describes some unusual lapidary projects utilizing fluorescent minerals and gemstones.

Chapter III: The Fluorescent Mineral Hobby

Richard C. Bostwick

The environment of the 1990s is far different from that described by Gleason for the period 40 years earlier. Expanding knowledge has deepened our appreciation for fluorescent minerals, but many localities once open to collectors have been built over, closed, or otherwise removed from access. In this chapter, a longtime hobbyist shares his secrets on how to get the most out of the hobby. He covers all the bases, from field collecting to displaying, and provides numerous insights into ways to acquire, study, preserve, and enjoy fluorescent minerals. He describes as well the many ways an enthusiastic collector can grow into other facets of the hobby, such as preserving mineral localities, lecturing about fluorescent minerals, and displaying at mineral shows.

Chapter IV: Activators in Fluorescent Minerals

Earl R. Verbeek

This chapter presents an in-depth answer to the question of why some minerals fluoresce but others do not. The author, a research geologist with the U.S. Geological Survey, reveals how chemical impurities and slight imperfections of crystal structure can activate bright fluorescence in minerals that otherwise would show no visible response to ultraviolet light. Discussed also are fluorescence quenchers—the opposite of activators—and some of the major fluorescent minerals and mineral groups. Throughout the chapter readers are given a glimpse into the world of the scientist, as the author explains how much has been learned about mineral fluorescence in recent decades—and how much more remains to be discovered.

FOREWORD

When Thomas Warren founded the Ultra-Violet Products Company, Franklin Delano Roosevelt was being swept into office for the first time. That was 1932, and Tom was busy with research that led to the invention of the handheld ultraviolet lamp—the technological tool which created our fluorescent-mineral hobby.

It was Thomas Warren who guided the ultraviolet-lamp industry through its first major crisis—World War II. When war broke out the U.S. government slapped a moratorium on civilian use of steel and copper, two key components of ultraviolet lamps. Not easily discouraged, Tom went to Washington where he was able to negotiate a deal with the government to keep producing lamps for mineral prospecting, especially for scheelite, the main ore of tungsten. This agreement enabled the United States to become self-sufficient in the mining of tungsten, a crucial ingredient used to harden steel.

Over the years, Tom has remained committed to sharing his passion for and knowledge of fluorescent minerals. He has helped organize and has participated in hundreds of mineral shows in the United States and in Europe. Through his lectures, he has opened the eyes of thousands to the enjoyment of fluorescent minerals.

Today's mineral collectors and fluorescent-mineral hobbyists have acknowledged their debt to Tom, honoring him with numerous awards, including an annual award to be given in his name and honorary memberships in four mineral and gem societies.

Given this background—a 60-year career which has spanned the life of our hobby—it is fitting that Thomas Warren, with this book, continues to share the magic of his true love—the fluorescent-mineral hobby.

DON NEWSOME
Founder and Past President
Fluorescent Mineral Society

PREFACE

This book has been prepared to meet the often-expressed need for a broad and accurate introduction to the fluorescent-mineral hobby. The subject matter is intended to complement, rather than duplicate, other recent texts on mineral fluorescence and differs from them chiefly in offering information not only on the minerals themselves but also on the nature of ultraviolet light, the lamps we use to produce it, and useful techniques for collecting, learning about, and displaying fluorescent minerals. The agents that cause minerals to fluoresce and those that act to prevent it are explained as well. Some of this information has never before been presented in nontechnical fashion and thus may be of interest not only to beginning hobbyists (our intended audience), but also to more advanced collectors.

The Magic of Ultraviolet Light

Thomas S. Warren

Enjoying Mineral Fluorescence

The world of fluorescence seems truly magical. The invisible rays from an ultraviolet lamp can cause a drab rock to explode in a rainbow of brilliant color. When you witness this incredible process for the first time a whole series of questions may enter your mind: What is happening? Why this sudden change? How does this magic work? The purpose of this book is to answer these questions.

Nowadays, fluorescence has become part of our everyday lives. We see it in the DayGlo™ colors of a child's sweatshirt, in the scary "black light" rides of an amusement park, in the invisible stamp which lets the nightclub bouncer know we've paid our admission. Man-made fluorescent dyes, paints, and inks, however, can't compete with nature. They can't match the thrill of watching one of your own seemingly ordinary rocks flare into fluorescent beauty. That thrill is probably what drew you to our hobby in the first place, and you'll find it only increases with years of collecting.

I stumbled upon the magic of mineral fluorescence nearly 60 years ago, and it has fascinated me ever since. Fluorescence technology was then very new. I was working to develop ultraviolet lamps for use in the medical sciences and any other market that offered even a limited potential. One market area was suggested by this country's need for the strategic metal tungsten, found in scheelite, an ore mineral which fluoresces bright blue. My colleagues and I believed that if a portable ultraviolet lamp could be made practical, then perhaps prospectors could find enough ore to supply the tungsten needed for military armaments required by the mobilization for World War II.

I mentioned this possible market to my distributor in New York. As luck would have it, he had a former college classmate who was then president of the New Jersey Zinc Company. This company owned the zinc mine in Franklin, New Jersey, and the mine operators knew that many of the minerals from the mine were fluorescent. At that time they used primitive equipment—a cumbersome and noisy electric arc—to create ultraviolet light in the laboratory in order to test some of the minerals. They didn't realize that the beautiful fluorescent minerals they were examining would turn out to be the world's most spectacular. My distributor told his friend of my work with ultraviolet light and, without informing me, asked his friend to send me some specimens from the mine.

A few weeks later a barrel of rocks arrived at my office. My surprise soon turned to amazement when I beamed my filtered ultraviolet lamp on the willemite and calcite rocks—they burst into brilliant green and red gems. This thrilling experience gave me a new interest. Throughout the following years, as I continued to refine fluorescence technology, I always kept the hobbyist in mind. I wanted him or her to be able to enjoy the magic of the fluorescence that had so captured my interest and imagination.

Although anyone can appreciate the beauty of a fluorescent mineral specimen on a purely visual level, understanding what makes it happen deepens the hobbyist's appreciation for his or her colorful collection. To understand fluorescence, however, we must first understand light.

Wavelengths of Light

The measuring of distances is second nature to most of us. Run a mile, walk a block. A yard—a foot—an inch: all are units of measurement. None of these, however, is practical as a unit of measurement for light; instead, we need a unit very much shorter. That unit is called the nanometer (abbreviated nm).

A nanometer is one millionth of a millimeter. Since one millimeter is about 1⁄25 of an inch, you will agree that a nanometer is a very short unit of measurement—so short that most of us cannot stretch our imagination enough to visualize it. However, as we become accustomed to using the term we will be able to see the importance of this tiny unit, for in some minerals a variation of only a few nanometers in the wavelength of the ultraviolet light beamed on the specimen can mean the difference between brilliant fluorescence and none at all.

A useful starting point for understanding light and its relation to fluorescence is the light that is most familiar to us all: sunlight. The electromagnetic energy that we receive from the sun covers a very broad range of wavelengths, of which visible light is only a small part. As shown in illustration no. 1, the human eye can see light only within a fairly narrow range of wavelengths, from about 400 nm to 700 nm. When our eyes receive light of these wavelengths, our brains interpret it in terms of color. All the colors of the rainbow lie within this wavelength range, from violet at 400 nm to red at 700 nm. To

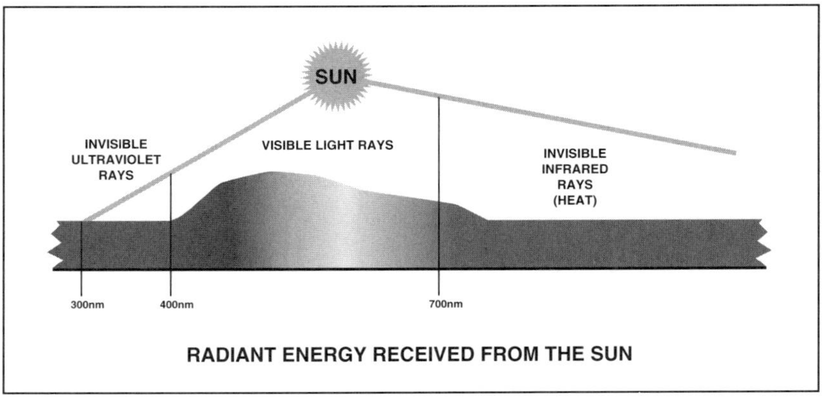

Illustration 1

either side of this wavelength range, however, lies a much broader realm of invisible electromagnetic energy.

Light with wavelengths longer than 700 nm, beyond the red part of the visible spectrum, is termed *infrared light.* We cannot see this light, but we can sense it as heat. The profound effect of infrared light on our everyday behavior is most obvious in summer, when commonly we attempt to avoid our exposure to it: we rest in the shade or use protective clothing to avoid becoming overheated or dehydrated.

Light with wavelengths shorter than 400 nm, below the violet part of the visible spectrum, is termed *ultraviolet light.* We can neither see nor feel this light, but that does not mean it has no effect on us. It is the ultraviolet component of sunlight, for example, that causes our skin to sunburn and tan. The biologic effects of ultraviolet light depend strongly on wavelength, as discussed below. Moreover, this light, as well as ultraviolet light of wavelengths shorter than those present in sunlight, has a variety of industrial uses, a few of which will be explored later in this chapter.

For general descriptive purposes, scientists have divided the wavelength range of ultraviolet light into four regions, each shown by separate shades in illustrations nos. 2, 3, and 4. A few of the many features of each of these regions will be explained below so that you can appreciate their importance in our lives and our hobby.

Light with wavelengths between 350 nm and 400 nm is called *longwave ultraviolet light* but is probably better known among hobbyists as "black light." It is present in sunlight and causes fluorescence in some minerals, notably fluorite, willemite, diamond and ruby. It seems to have no deleterious effect on either skin or body. "Black light" is widely used by hobbyists because of the ready availability of BLB ("black-light-blue") tubes, which provide an inexpensive means of producing invisible ultraviolet light with wavelengths around 360 nm. These tubes are popular with young people because they cause fluorescence in many common household items. Teeth, eyes, fingernails, and skin blemishes also fluoresce, in addition to many paints and some fabrics. For mineral fluorescence, however, the BLB tube is less than ideal. The visible blue light which BLB tubes emit along with the ultraviolet light overcomes the fluorescence of many minerals and distorts the true color of fluorescence in all of them. For this reason, an ultraviolet lamp with a separate filter to screen out visible light is recommended for the mineral hobbyist.

Also present in sunlight is *midwave ultraviolet light,* with wavelengths between 300 nm and 350 nm. Midwave ultraviolet light produces an interesting fluorescence in some minerals that respond less well to ultraviolet light of shorter or longer wavelength. I have found some eucryptites, agrellite from Quebec, and a few calcites from the Castle Dome Mountains of Arizona to be especially intriguing. But the effect of midwave ultraviolet light varies greatly among different minerals. Most scheelite, for example, does not react; the minerals from Franklin, N.J., are variable in responding; and some of the

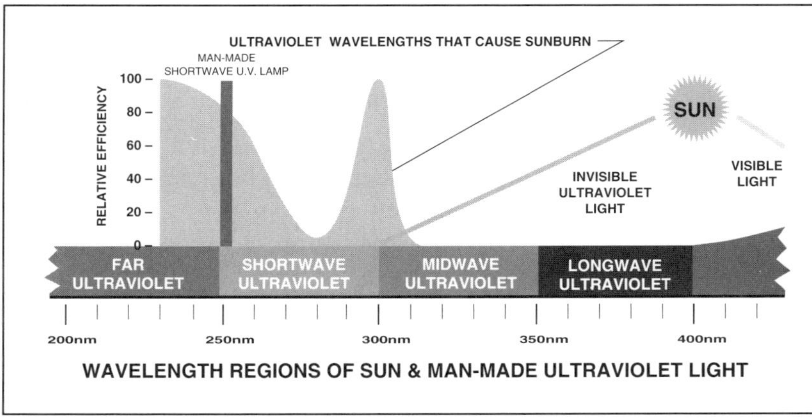

Illustration 2

scapolites do very well.

Ultraviolet light near the lower wavelength limit of the midwave region, around 300 nm, is markedly actinic—that is, it induces photochemical changes in the human body. Some of these actinic effects are beneficial and others harmful, depending primarily on length of exposure. Modest exposure assists the body in absorbing vitamin D and also causes skin to tan, but overexposure can result in painful sunburn and increased risk of skin cancer. The odds of overexposure are greatest at high altitudes, where the thinness of the Earth's atmosphere affords lesser protection from sunlight, but they are also increased during the summer months when the sun is at its highest. A related but beneficial actinic effect is that light of about 300 nm wavelength is fatal to many bacteria and can thus be used to create and maintain a sterile environment.

Shortwave ultraviolet light, with wavelengths between 250 nm and 300 nm, is not present in sunlight because it is absorbed by the Earth's atmosphere. Light in this wavelength range is remarkably effective in inducing fluorescence in minerals, and lamps that produce it are thus widely used by hobbyists. The range of fluorescent effects and the types of shortwave ultraviolet lamps currently available will be discussed more fully later in this chapter. Like light of slightly longer wavelength, shortwave ultraviolet light efficiently stimulates vitamin D activity in the human body and is bactericidal. Severe overexposure to light of this wavelength can cause sunburn (but not a tan), but this is hardly a cause of concern to the hobbyist—the low-wattage lamps conventionally used for mineral fluorescence are no more harmful than a color TV when used properly.

The wavelength range of the fourth region of ultraviolet light, termed *far ultraviolet light* or *vacuum ultraviolet light,* extends from 25 nm to 250 nm and is shown on illustrations nos. 2, 3 and 4. Far ultraviolet light is effectively absorbed by air and thus, like shortwave ultraviolet light, is not present in sunlight which reaches the Earth's surface. Most research on far ultraviolet light must be conducted in vacuum chambers, and consequently we are not concerned with it in the mineral-collecting hobby.

This review of the various wavelengths of ultraviolet light is presented with the expectation that the more we know about the components of our hobby, the more we will enjoy it. I know you will find that true of the other chapters in this book as well.

Some of the commercial and industrial uses for ultraviolet light are discussed in the following pages. Only a select few of the applications are mentioned, but they are typical of the wide use being made today of this energetic light.

In summary, only ultraviolet light with wavelengths shorter than 310 nm will cause sunburn, Vitamin D activity, and bacteracidal action. The same light stimulates the brightest fluorescence in most fluorescent minerals. Light of longer wavelength will not cause these effects on the human body and for most minerals is less effective in inducing fluorescence, as described later in this book.

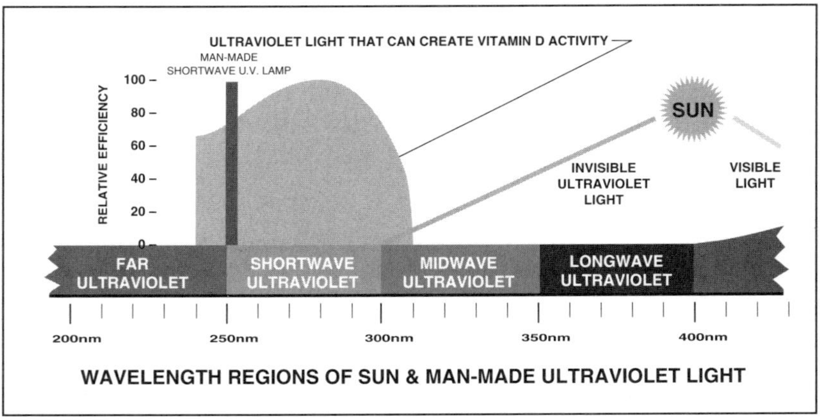

Illustration 3

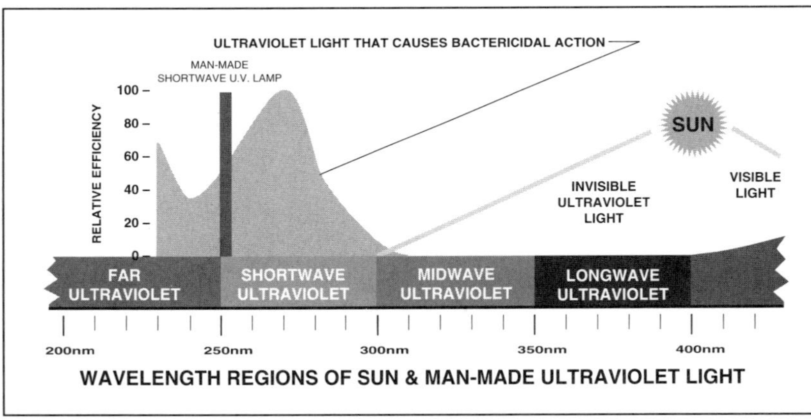

Illustration 4

The Many Uses of Ultraviolet Light

Scientific minds have long explored ways to expand the usefulness of ultraviolet energy. We do not know where the next practical idea will appear. It is possible that you, the reader, will have some specific idea that will open a whole new industrial or commercial use for ultraviolet light.

Currently, longwave, midwave, and shortwave ultraviolet light are widely used in numerous laboratory applications and processes. Today's growing emphasis on the study of human, animal, and plant genetics has greatly increased the market for ultraviolet devices for viewing, photographing, and crosslinking of DNA/RNA electrophoresis gels. Ultraviolet light is routinely used in both the thin-layer chromatography and paper chromatography processes and is extremely important in the study and analysis of cell cultures and biological samples.

In the medical field, ultraviolet light is used in the treatment of many illnesses, particularly skin disorders. Ultraviolet light also has applications in ophthalmology (the treatment of eye problems) and in surgical procedures. Many hospitals employ ultraviolet lamps for the destruction of bacteria to create a sterile atmosphere.

The ability of this invisible ultraviolet energy to kill bacteria has created other markets for shortwave ultraviolet light in baking, food processing, cosmetic manufacturing, water purification, and, most recently, the treatment of hazardous waste.

In the agricultural and food industries, ultraviolet light is widely used as part of the quality control process, to detect naturally fluorescent contaminants that could present possible health hazards. Stored food and supplies can be examined for signs of rodent urine, which fluoresces. Grain and peanuts can be checked for the presence of the naturally fluorescent aflatoxin mold, a carcinogen.

Ultraviolet light also has a major quality control application in the field of nondestructive testing. It is used in foundries and in aerospace and defense factories to detect stress and cracks in metals.

Ultraviolet light has been used for many years in the financial and security fields to detect counterfeit bills and to help identify stolen property that has been marked with brightly fluorescent invisible ink or powder. Savings and loan institutions have used a system of verification that allows a signature to be transferred invisibly to a passbook. The signature becomes visible when the document is placed under an ultraviolet lamp. The areas of criminology and forensics also employ many techniques requiring ultraviolet light.

As you can see, uses for ultraviolet light are plentiful and continually expanding. One clever professor called upon his knowledge of ultraviolet light to keep his students on the "up and up." In the course of a school year, students were given the task of collecting a variety of insect specimens. At the end of the year, the professor marked the specimens with an invisible fluorescent spray. In this way he was able to identify students who merely recycled old specimens instead of collecting their own.

Some uses of ultraviolet light, however, are not so beneficial. Many people wish to place fluorescent minerals in, and an ultraviolet light over, their fish aquarium. This should be discouraged for several reasons, the most important being that fish have no eyelids and the constant exposure to shortwave ultraviolet light can damage their eyes.

Ultraviolet Light and Our Eyes

The natural curiosity which we humans possess is often a good thing. After all, it is curiosity about the strange glowing rocks which gave birth to our hobby. But the limits of exploration need to be drawn at a safe level. For collectors of fluorescent minerals, that means heeding the written warning on the ultraviolet lamp—"don't look at the shortwave ultraviolet light."

This warning often creates a feeling of fear in some people when viewing fluorescent minerals. Although there is no reason for fear, there *is* reason for caution—no one wants to get sunburned eyes. Like a sunburn of the skin, sunburn of the eyes is temporary, not permanent. The long-term effects, however, are still unknown. As with sunburn of the skin, sunburned eyes are not caused by a brief exposure to ultraviolet light. A short, superficial look will normally have no effect on the eyes, except that with some people it will cause the eyes to fluoresce, creating a momentary sense of haze. In any case, looking at the light source is a pointless activity—the ultraviolet rays are invisible, so you can't see them regardless of how long you look.

If you do look into the lighted shortwave lamp for a long enough time, however, you *will* get sunburned eyes. This can be a miserable experience and justifies the warning. The full effect of this sunburn becomes apparent six or eight hours after exposure. The eyes get weepy and feel like they have sand in them when you blink. It is very frightening until you know it is only temporary. If you go to a doctor, the usual prescription is a mild anesthetic that will help ease the gritty feeling. In all cases, the best remedy is time. The sunburn typically fades after a few hours of misery, and over the years I have heard of no cases of permanent damage. Still, a mineral enthusiast can avoid this uncomfortable effect simply by exercising caution.

If you wear glasses, they should provide complete protection on direct viewing, but be cautious of rays that can reach the eyes from a lamp shining toward you from the side while you are looking ahead.

Be cautious, too, of ultraviolet rays reflected from the specimens you are examining. Most minerals absorb ultraviolet rays but others reflect them, particularly when polished, and so your eyes may be exposed to ultraviolet light even if the lamp is never pointed in your direction. Glasses or protective goggles will protect you from this reflected light and should be used whenever you plan to examine fluorescent minerals for a prolonged period of time.

All of the above is in reference to the shortwave ultraviolet rays. When longwave or "black light" lamps are being used, the eyes will fluoresce, but these lamps do not cause sunburn.

As with most things in our modern world, exercising common sense with ultraviolet light is best. You wouldn't shine a flashlight directly into a person's eyes or place your hand through a laser beam for extended periods of time, so why treat ultraviolet light this way? The hobbyist understands this—his or her commitment to using ultraviolet light safely will ensure years and years of enjoyment of this fascinating hobby.

To increase this enjoyment, of course, it helps to have a good understanding of the central star in the ultraviolet universe—the lamp.

Fluorescent and Ultraviolet Lamps

We can learn a lot about the ultraviolet lamps which are so crucial to our hobby by looking at their consumer cousins—the brilliant white fluorescent tubes which light up our offices, restaurants and parking lots. Like ultraviolet lamps, fluorescent lamps are constructed from glass tubes with cathodes inserted at either end. During manufacture of a fluorescent lamp the tube is coated with a phosphor on the inside walls, a vacuum is created in the tube and argon gas is inserted along with a small drop of mercury. A ballast controls the voltage. When the lamp is turned on, the current between the cathodes vaporizes the mercury and causes the mercury vapor to emit copious quantities of ultraviolet light. This light is absorbed by the phosphor coating which lines the tube, and is re-emitted as visible light. These lamps require only a small input of electricity and can be made to provide fluorescent light in any color of the rainbow.

There are two main differences between fluorescent lamps for illumination and the ultraviolet lamps used in our hobby. First, most of the light produced by ultraviolet lamps is invisible. Second, shortwave ultraviolet rays will not pass through most types of glass or most plastics. This limits the types of materials that can be used to manufacture the tubes for shortwave ultraviolet lamps. Currently, shortwave ultraviolet lamp tubes must be made from quartz or high-silica glass.

The two main types of ultraviolet lamps used for mineral fluorescence are the longwave and the shortwave. The shortwave ultraviolet lamp uses a clear tube, while the longwave ultraviolet lamp uses a tube with a longwave-emitting phosphor coating on the inside. The tubes for longwave lamps are similar in appearance and construction to conventional fluorescent tubes, but the phosphor coating is designed to emit longwave ultraviolet light rather than visible light.

All fluorescent lamps are extremely efficient. This efficiency, discussed below, is a key reason for the widespread popularity of fluorescent lighting.

The Amazing Ultraviolet Intensity

In shortwave ultraviolet lamps, mercury atoms perform a double duty called *mercury resonance radiation.* Because of this property, ultraviolet lamps are actually producing two to three times the amount of ultraviolet rays as a conventional lamp would produce visible rays. A 40-watt or 60-watt incandescent light bulb is not a very strong light source for home use, yet a 4-watt or 6-watt ultraviolet lamp has enough energy to create bright fluorescence in a small area, and a 15-watt ultraviolet lamp has enough energy to cause an average mineral display to fluoresce brightly. Due to the tremendous efficiency of the mercury discharge, which releases most of its energy at a wavelength of 254 nm *(illustration no. 5),* hobbyists will find the lamps inexpensive to operate and very effective for mineral fluorescence.

This extra-strong energy level consequently makes it possible to manufacture a lamp that is extremely efficient in emit-

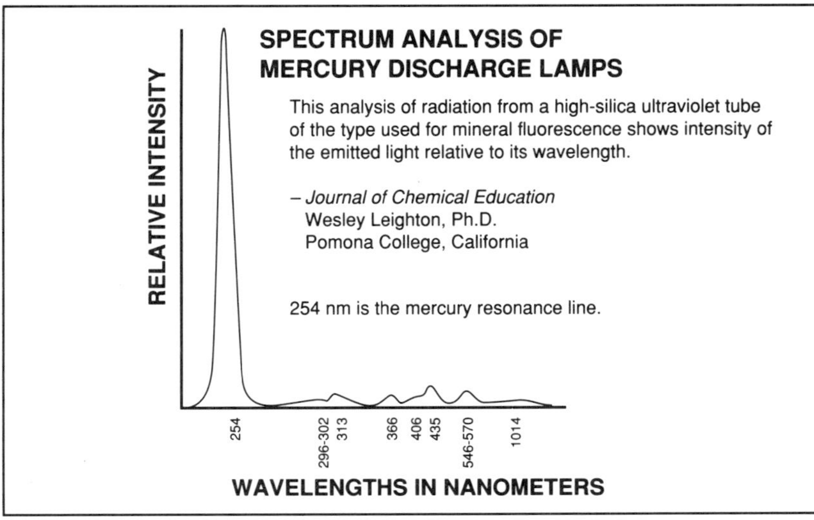

Illustration 5

ting ultraviolet light and very poor in emitting visible light, but this is exactly what the hobbyist requires. Broken down into numbers, 89.2% of the energy from a shortwave tube is emitted at a wavelength of 254 nm, while only 3.8% spills over into the other ultraviolet wavelengths *(illustration no. 5)*. Less than 7% is emitted as visible light and heat. However, even this small amount of visible light must be eliminated as much as possible, and this is done by the special filters we use on our lamps.

Filters

The filters on our ultraviolet lamps are just as important as the light source itself. Without a filter, the visible light from the lamp would overwhelm the fluorescence of most minerals. With a filter, in contrast, the visible light is effectively absorbed and the ultraviolet light allowed to pass through. Some people have been under the impression that the filters are responsible for the ultraviolet light. They have placed a blue filter over a flashlight and tried to make a prospecting lamp. This does not work; changing the color of visible light does not transform it into ultraviolet light. The filters for shortwave ultraviolet lamps are specially designed for the mineral hobby and are actually more difficult to make than the rest of the lamp.

These filters, seemingly almost black but actually dark purple in transmitted light, were originally made from a phosphate glass whose composition is wholly unlike that of conventional household glass. The composition of this shortwave filter glass and the way in which it is manufactured have changed over the years. These changes have affected our hobby because hobbyists had to adjust time and again to the introduction of filters of varying quality.

The shortwave filters made from 1930 to 1960 did very well when first used, but their main problem was that the shortwave ultraviolet rays caused a slow physical and chemical change, called solarization, in the glass. This resulted in a gradually increasing absorption of ultraviolet light by the filter so that after a few hundred hours of use the filter was worn out. After 1960 a change in filter formula was made which increased this problem rather than correcting it. The result during the 1970s was that shortwave filters absorbed both visible light and a too-large proportion of ultraviolet light. This effect, coupled with solarization, caused many collectors to lose interest in the hobby.

A dramatic change came in 1980 when a new shortwave filter came on the market. The manufacturer apparently used a new technology in formulating the filter because it not only had a high transmission of shortwave ultraviolet rays but also had an extremely low rate of solarization. In testing several 4-watt lamps I found that the filter was almost as good after 6,000 hours as at the start of my test. In fact, the light source itself—the tube of high-silica glass within which the ultraviolet light was generated—solarized more quickly than did the filter. Later tests showed, not surprisingly, that higher-intensity sources of ultraviolet rays increased the rate of solarization of shortwave filters more rapidly than the low-intensity lamps (of 4 watts or 6 watts)—but still the new filters remained usable for hundreds of hours longer than any filter made before.

At the present time, only one company makes this new type of filter, but several others around the world are trying to duplicate the results. When a comparable or better filter is made it may bring the price down. Currently the filter is by far the most costly component of shortwave ultraviolet lamps, and thus their manufacture today represents a compromise between cost and efficiency. Should the cost of shortwave filters decrease, larger filters can be used to make lamps more efficient.

The filter in a longwave ultraviolet lamp is entirely different because, unlike shortwave ultraviolet rays, longwave rays *will* pass through most types of glass. A filter which will screen out visible light and pass longwave ultraviolet light will not solarize; thus the effectiveness of such filters does not decrease with time. Several companies make these longwave filters, and prices remain stable and reasonable.

In one type of longwave ultraviolet lamp the tube itself is made of filter glass; this self-filtering unit is known as a "black-light-blue" tube. This is designed for theatrical effects with fluorescent paints and fabrics but is not suitable for minerals because it allows too much blue light to pass through the filter, thereby altering the perceived color of fluorescence in most minerals. For best results with longwave fluorescence, hobbyists should use a lamp with a separate tube and filter.

Selecting an Ultraviolet Lamp

For the hobbyist interested in fluorescent minerals there are two types of lamps from which to choose—portable battery-operated lamps for field trips, and lamps for home use which operate on regular household current. Both types can be purchased to produce either longwave or shortwave ultraviolet light, or in some models both.

Currently, battery-operated lamps come in two strengths—4 watts and 6 watts. The 4-watt lamp is designed for beginners and those with only an occasional need for an ultraviolet lamp. Originally these 4-watt lamps were made in the Orient as flashlights and were fitted with small fluorescent tubes similar to those used in household lighting. Manufacturers of ultraviolet lamps, however, adapted this lamp to create longwave or shortwave ultraviolet light by the use of longwave tubes or shortwave ultraviolet tubes and filters.

To convert the lamp to a longwave ultraviolet source, the white-light tube is simply replaced with a "black-light-blue" tube. While this is easy to do, the amount of longwave ultraviolet light produced is small and the blue glass filter does not absorb enough visible light to make a good lamp for mineral fluorescence.

Alternatively, a battery-powered 4-watt lamp can be converted to a shortwave ultraviolet lamp. To do this, the original tube is replaced with a shortwave ultraviolet tube of the same size, and a shortwave filter is placed over the tube to exclude visible light. The conversion provides a usable lamp, but not one that will result in bright fluorescence except at very close range. Such a lamp is satisfactory for examining single specimens in a well-darkened area, but for general prospecting a much stronger light source is needed. In summary, the 4-watt battery-powered lamp is fine for beginners due to its low cost, but most hobbyists will soon wish for a more powerful lamp.

The second type of battery-operated lamp is the 6-watt unit. It is available with either one or two tubes and operates with standard lantern batteries weighing 1½ lbs. each. This type of lamp is a great deal stronger than the smaller 4-watt unit. On a dark night, as one walks along with the lamp directed downward, the fluorescence of many minerals will be evident without the need to stoop to bring the lamp closer to the ground. Some of these lamps can be operated both on battery and on home current. They are a good option for a person who plans to spend a large amount of time hunting fluorescent minerals in the field.

The hobbyist should realize that the source of power for a battery-operated lamp is direct current. This means that the electrical current flows in one direction at all times. Because of this, one of the filaments in the tube will start to wear out and deposit a black coating on the wall of the tube. This drastically shortens the life of the tube. When the light from the tube starts to dim, a temporary solution is to reverse the tube in the lamp. It will then continue to operate for a longer period of time.

As with their prospecting cousins, lamps for home use are available in various sizes and intensities, all operating on alternating current from ordinary wall outlets. Most popular here in the United States have been the 6-watt handheld lamps. A 4-watt lamp is also available, but it is not sufficiently strong to bring out the full beauty of most fluorescent minerals.

Lamps using either one or two 15-watt tubes have been gaining in popularity. Such lamps can stimulate much brighter fluorescence in minerals than 4-watt and 6-watt lamps, but there is a drawback: the 15-watt tubes are greater in length and diameter than those of lower wattage, and thus a larger filter is needed. Inasmuch as the filter is the single most expensive part of an ultraviolet lamp, the 15-watt lamps are considerably more costly than smaller varieties.

All of the lamps mentioned contain a polished aluminum reflector mounted behind the tube. The circular tube radiates light in all directions, some through the filter but much of it into the lamp housing. The purpose of the reflector is to redirect some of this light back through the filter and thus increase the intensity of the ultraviolet beam. Generally the filter is two to three times the width of the tube, but in view of the high cost of filter glass it is becoming common to employ filters only as wide as the tube itself, which reduces their effectiveness; for all practical purposes the reflected ultraviolet rays are lost, though some may incidentally be transmitted back through the tube. Still, by sacrificing this component of the ultraviolet light, a good lamp can be produced at a reasonable price. Because the loss of efficiency is slight compared to the considerable reduction in cost, the trend toward narrow filter designs seems sure to continue.

Currently on the technological horizon is a 25-watt lamp using the "narrow lamp" configuration. This lamp will be able to stimulate much brighter fluorescence in minerals than the 15-watt variety with only a small increase in price.

While we can't see into the future, progress continues to be made. We can only hope that this will lead to better and better ultraviolet lamps, allowing greater enjoyment of fluorescent minerals.

Multiband Lamps

Collectors often feel torn between buying a shortwave or longwave lamp. A longwave lamp is the less expensive of the two because both the tube and the filter are less costly than the comparable items for shortwave lamps. The major drawback of a longwave lamp, however, is that about 90 percent of known fluorescent minerals fluoresce better under shortwave ultraviolet light—thus the money saved is offset by considerable loss of opportunity to appreciate some of the world's most interesting fluorescent minerals. The obvious solution is to purchase both types of lamps, but for some collectors the cost can be prohibitive. For these people a multiband lamp, one that produces both shortwave and longwave ultraviolet light from a single unit, may provide a happy alternative.

A multiband lamp is made by coating half of the inside of a shortwave tube with a longwave phosphor. Many people like this arrangement for its convenience, but by sacrificing half the shortwave energy for conversion to longwave ultraviolet light, the brightness of fluorescence of most minerals is diminished. For nighttime collecting, the multiband lamp may not provide sufficient fluorescent brightness to avoid overlooking some fluorescent minerals. In the home, the diminished intensity of a multiband lamp can be partially overcome by simply holding the lamp closer to the specimens. Still, inasmuch as brilliance of fluorescence adds greatly to our pleasure in viewing fluorescent minerals, the purchase of separate longwave and shortwave lamps rather than a single multiband unit is worth the additional cost.

Cold Quartz Display Lamp

There is another type of shortwave lamp that deserves attention—the cold quartz display lamp. While most display lamps use 15-watt germicidal ultraviolet tubes, the cold quartz display lamp uses a fused quartz grid. There are several advantages to this quartz lamp. Not only will the tube last much longer, but the ultraviolet light produced by a fused quartz lamp is of much higher intensity than that emitted from standard germicidal lamps. The cold quartz lamp will turn on or off instantly without shortening its operating life. It also uses a high-voltage transformer rather than a low-voltage ballast. It operates on the same basis as the common neon sign except it uses quartz tubing instead of normal glass. The main drawback to this lamp, however, is its high cost.

Wherever frequent on-off demonstrations are required, as in museum displays, a regular 15-watt tube will often hesitate when restarting. This can be a real problem, as it both distracts people and shortens the life of the tube. The regular 15-watt tubes also gradually solarize and lose some of their ability to emit ultraviolet rays by the time they have been burning for a few hundred hours. The quartz grid lamps, in contrast, can operate for 10,000 or more hours with little lessening of ultraviolet output. These features may or may not justify the difference in cost. Both types of lamps use the same filter.

How Minerals Fluoresce

Mineral fluorescence is beautiful and fascinating, and it becomes even more interesting when we understand some of the science behind it. To do this we must understand something of the atomic structure of minerals and how the atoms which make up minerals respond to ultraviolet light. Atoms are the basic building blocks of all matter, and atoms themselves are made up of a small, dense nucleus in the center with unimaginably tiny electrons orbiting relatively far away. You can think of an atom as much like our Solar System, with the Sun in the center and the planets—Earth, Jupiter, Saturn, and so on—in more or less circular orbits far away. In fact, the structure of the atom is far more complex than this, but for the purposes of discussion, this analogy will suffice.

Of the three general classes of matter—gas, liquid, and solid—it is the latter that most interests us from the point of view of mineral fluorescence. In most solids, and nearly all minerals, the atoms are arranged in a precise, three-dimensional array. Minerals for which this is true are said to be *crystalline*. As minerals crystallize in the Earth, it is common for "stray" atoms to become incorporated in the crystal structure. These impurities are important, because sometimes they

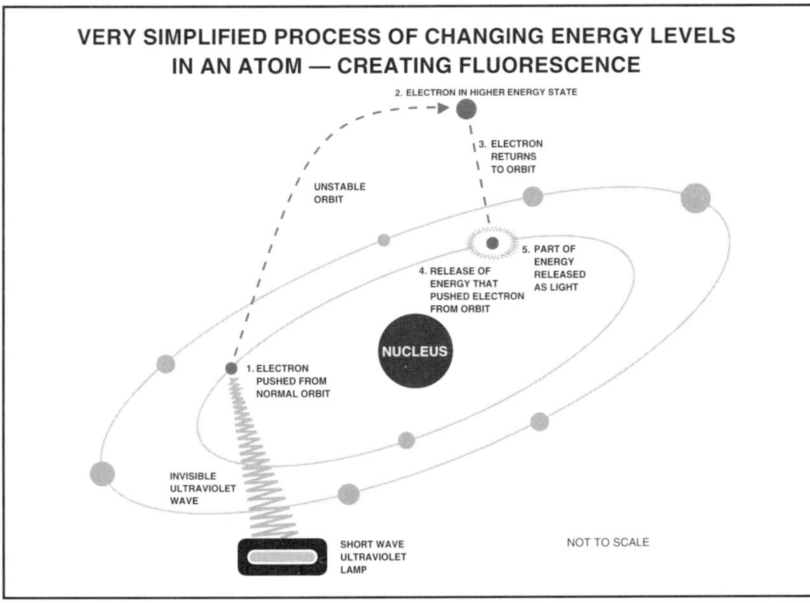

Illustration 6

cause a mineral that would not ordinarily fluoresce to do so. Such impurities are called *activators*. Depending on which activators have become incorporated into its crystal structure, a given mineral can show differing colors of fluorescence. For example, pure calcite will not fluoresce. But add a little manganese, and you may get a brilliant red fluorescence. Add a little uranium, and the fluorescence may be green. The various ways in which activators cause minerals to fluoresce are discussed in greater length in a later chapter of this book.

We've already discussed the nature of ultraviolet light—how its wavelength range lies below that of light visible to the human eye. But while our eyes cannot detect ultraviolet rays, the electrons in many atoms can absorb them. When an electron orbiting a nucleus absorbs the energy of some ultraviolet light, it becomes more energetic, and moves into a higher orbit *(illustration no. 6)*. But this higher orbit is unstable, and the electron eventually has to give up this extra energy to resume its old position. A scientist would say that the electron is returning to its original ground state.

If the atom released its extra energy all at once, it would emit the same ultraviolet light that activated it. But atoms commonly release their extra energy in stages. Some of the extra energy is transformed not into light but into vibrational energy, which we experience as heat. (Ordinarily, in the case of fluorescent minerals, the amount of heat released is far too small to be detected without special instrumentation.) When some of the energy is released as heat, the rest of the energy can be released as light. But this light will then be less energetic—at a longer wavelength—than the original ultraviolet light. If the wavelength of the emitted light is long enough, it will come into the visible range and we will be able to see it as glowing color. This is fluorescence.

How to Display Your Specimens

While mineral collections can range in size from several specimens to thousands, there is one problem which is shared by all collectors. The bright fluorescence you want your visitors to see can be appreciated only in a dimly lit room. The fluorescence is overwhelmed in a brightly lit room. If you are showing your collection of minerals at night, the problem is easy to solve. Have a small light in the opposite corner of the room or in an adjacent room with the door open. A small amount of light is necessary to feel comfortable and to see your way around, but your specimens will still glow nicely because they have no competing direct light. If you show your display in the daytime, be sure to close the window curtains or drapes and have your collection at the opposite wall so that your visitors will have their backs to the window. Collections can be arranged on the shelves of a bookcase or in the drawers of a cabinet, ready to show at a moment's notice. However, if you live in small quarters, it may be necessary to keep your minerals in boxes and, when your guests arrive, display them in a room with no windows such as a bathroom or large closet.

Whether you display your collection in a bookcase, drawer, or on a table top, one absolute requirement is that you use a black cloth for the background. It is also important that you have no sources of competing fluorescence around your display—no fluorescent paper, cotton, fabric, or any object that fluoresces.

With everything in place, you are ready to put on a dazzling display of fluorescent beauty. Have your ultraviolet lamp in hand and show off your specimens one by one. Point out any special feature that interests you about a mineral. Take your time. Your friends will have lots of questions, and a careful reading of this book should give you answers to many of them.

You can tell them what wavelengths of ultraviolet light you are using and explain the many industrial and commercial applications for ultraviolet light, as well as its beneficial health properties, and the unusual efficiency of the fluorescent lamps used everywhere for illumination. Even more important, you can explain about variations in fluorescence of minerals. You can point out how small variations in the composition of minerals can cause dramatic differences in fluorescent quality, and why some specimens of a particular mineral will fluoresce but others of the same mineral won't. Nature has millions of possible variations, and in many cases these can best be discovered with an ultraviolet lamp.

The magic of mineral fluorescence is a special feature that few people know about, and it is a delight to open friends' eyes to this fascinating new world. Your friends will appreciate both your enthusiasm and your knowledge of these minerals which have beauty beyond belief.

Thrills and Surprises

Our hobby is filled with thrills and surprises. One of the greatest of these is discovering a new fluorescent mineral or locating a new site for fluorescent specimens.

But such discoveries are usually made after a lot of footwork, and many people prefer to let someone else do the discovering. If you are one of these people, you can enjoy the fruits of many prospecting trips with a single visit to a mineral store, many of which stock fluorescent mineral specimens and lamps. If you prefer to order by mail, you may secure a fluorescent mineral catalog from such a store. There you will find a cornucopia of fluorescent minerals and lamps. In this way you can amass a collection of minerals which spans a geographic and historical range far greater than that which you can personally access.

If you do go prospecting and take your ultraviolet lamp into the southern deserts, you may find what appears to be a bright green fluorescent stone. As you reach for the specimen you may be startled when it suddenly scampers away. Lucky for you it does, for scorpions fluoresce and their sting is painful—and they really don't add much to a mineral collection.

Scorpions are rarely the goal of a fluorescent mineral hunt, but their presence does underscore the rock hunting credo—you never know what you will find. One amateur prospector who roamed the Panamint Mountains of Death Valley learned the truth of this statement, much to his delight. He had been camped at a gold-mining area for about a week, prospecting every night with his shortwave ultraviolet lamp. On his last night before returning to the big city, he was hoping to find the signs of scheelite, the brilliant blue-fluorescing ore of tungsten.

He explored new territory with no success. Dejected, he returned to camp before daylight with his lamp shining on the ground as he walked; he knew he might as well use up the batteries. Suddenly he saw a glow. In this haphazard way he uncovered a treasure trove.

The next morning the prospector showed samples of his discovery to the mine foreman, who immediately recognized them as scheelite. "Where did you find it?" the foreman asked. "Could you find your way back to the location?" The answer was "Yes"—it was just a few hundred yards away. The find turned out to be on the mining company property. What the prospector had stumbled onto turned into a lucrative mining venture.

You, too, may discover hidden and interesting values in finding fluorescent minerals.

Conclusion

Our hobby is always changing. As new technology becomes available, the methods we use to discover and display our fluorescent treasures continue to improve. The fascination for glowing minerals, however, remains constant—it is as intense in hobbyists today as it was 60 years ago. As we break down the process of fluorescence and look at its various components we see that, indeed, this phenomenon is not *really* magic. But as we watch the glowing transformation of an ordinary rock by ultraviolet light, we realize that, magic or not, fluorescence holds a fascination for us which never dims.

We are seeing things as we have never seen them before. It has been suggested that you "look at the world through the eyes of child" if you want to see *something differently.* We, as hobbyists, use the magic of ultraviolet light to see some things—minerals—differently, as we now have a second pair of eyes. In this chapter I have explained how this second pair of eyes enables you to see things you have never seen before. You now know what happens and why it happens, so the next time you go on a weekend trip or vacation, be sure to take an extra pair of eyes with you. You never know what you may be walking over. Hidden values may be under your feet.

Acknowledgements

In closing, I wish to express my appreciation for the work done by Dr. Phillip Leighton in preparing the information contained in illustrations nos. 2, 3, and 4. Dr. Leighton furnished the information to me while he was professor of Chemistry and Photochemistry and the executive head of the Department of Chemistry at Stanford University, Palo Alto, California. He expressed the opinion that this information is basic for the study of ultraviolet light on living cells as well as for the fluorescence of minerals. Also my appreciation to Dr. Wesley Leighton of Pomona College, Claremont, California, for illustration no. 5. For further reference, two recent books provide interesting reading on ultraviolet rays: *Light—Medicine of the Future,* by Jacob Liberman, O.D., Ph.D., published by Bear and Co., Santa Fe, New Mexico 87504; and *Sunlight,* by Zane R. Kime, M.D., M.S., published by World Health Publications, Penryn, California 95663. In addition, the comments and additions by the following friends have been most helpful—Carolyn Cory, Marilyn Grout, Jessie Ramirez, Don Newsome, Don Nelson, and others.

The Fascination of Fluorescent Minerals

Sterling Gleason

This chapter consists of excerpts from Sterling Gleason's classic Ultraviolet Guide to Minerals, *published in 1960. Many of his thoughtful and enthusiastic observations remain timely, as you will discover. A few minor changes have been made to update the text and integrate it in this book, but what follows is essentially vintage Gleason.*

First Experiences

The world of fluorescence is a world of beautiful color. In the darkness all the ordinary colors of our daylight world disappear. Only the intensely glowing hues of fluorescent substances touched by the ultraviolet beam shine out with striking clarity. It is this selective ability to awaken new radiance in certain materials that makes the ultraviolet lamp so useful to the mineral hunter.

Let's get acquainted this evening with our new lamp by taking an imaginary trip under the stars.

Before we step out, unpack the lamp and snap on the beam. The spooky off-white tint of your fingernails, the livid purplish reflection from your hands, look ghostly. A violet streak on your palm is scar tissue of an old cut, long healed. Reappearing also are a number of small freckles, almost invisible in daylight. Between the index and second fingers of your hand, a smudgy area glows reddish-orange—the fluorescent mark of the habitual smoker.

Your white shirt is a dazzling expanse of luminous bluish-white, the brilliant sheen produced by detergents used in laundering. Those mysterious, Chinese-looking inscriptions are fluorescent laundry marks.

Flashing the lamp about the room reveals little in the dim purplish light that comes through the filter. Ultraviolet light is, of course, invisible. A line of printing on a wall calendar leaps into orange brilliance, for many inks are fluorescent. A plastic dish on the table, catching the ultraviolet rays, gleams brightly pink. Nearby there is a flash of brilliant green: a string of costume jewelry, dime-store variety but strikingly fluorescent.

Stepping outside, we notice bright blue-fluorescent spots on the driveway: a glowing film of oil drippings. (They could be yellow, orange, or pale green, depending on the dye used to give "bloom" to the oil.) Some auto mechanics use the lamp to determine if a leak comes from crankcase or automatic transmission. In a mine, the same glow, caused by flying oil drips thrown by machinery, often appears to be streaks of fluorescent ore. Amateur lapidaries sometimes are startled to see a blue haze on their specimens, due to kerosene in the cutting oil.

Where to Look at Night

We climb into the car and head for the nearest place where rocks abound, a city park where a stream passes through a narrow gap between rocky outcrops. As we squat on the bank, little is visible but the purple reflection of the lamp's visible light upon the water. Soon we make out bright yellow specks upon a boulder, just above the high-water line. They are lichens, soft, velvety, easily scratched off, fluorescent when dry.

Pebbles nearby show a creamy white glow: a coating of caliche, easily scratched by the fingernail, twin sister to the white scale in your teakettle. When this material, predominantly calcite, is deposited on rock surfaces by water, it dries and weathers to form a fluorescent coating. Its color in the lamp's rays may be creamy yellow or orange; other areas that have not weathered so much may be pale green or blue. The undersurface of pebbles may glow white or cream and almost invariably will be phosphorescent (usually for about five to fifteen seconds). Calcite is one of the most common of fluorescent minerals, and especially so is this weathered caliche.

Scooping up a handful of sand, you discover half a dozen tiny bright flecks like grains of gold: zircon, most likely, probably from miles away. Several minute grains wink brightly like blue-white diamonds. Their brilliance suggests scheelite *(plate 7)*, but no further signs of tungsten are seen.

Some of the dried grass blades light up in clear yellow-brown tones as the ultraviolet beam strikes them. You snap nearby branches and examine the broken ends for fluorescence. Certain woods, including locust, cypress, and sumac, can be turned and polished to make handsome fluorescent candlesticks, book ends, or other decorative objects.

Further exploration reveals nothing but a broken bottle, fluorescent brilliant yellow-green from uranium traces often found in glass. The rocky outcrops show no response. We decide to drive over to a nearby abandoned quarry.

No sooner have we scrambled down to path's end than our lamp picks up a brilliant green stripe outlining a seam in the rock. It is very smooth to the touch and the color seems to come from the very depths. Neither fingernail nor knife will scratch it. Hardness, color, slippery feel, all point in one direction—to the much-traveled silicas of the quartz family.

What You May Find

Agate, jasper, opal—these turn up in virtually every region. They fill fissures and cavities and coat rocks. They even penetrate into the hearts of fallen trees, replacing decaying wood, cell by cell, until whole forests have been converted into storehouses of gemlike petrified wood. You may recognize the wood fibers and even the bark. Fossilized bones likewise become agatized. Both may absorb minute traces of uranium that imparts a bright green fluorescence—but no phosphorescence.

Jasper almost never fluoresces; agate does sometimes, and opal quite often. Opal is slightly softer and its characteristic slick "opally" feel is almost enough to establish its identity. Jasper and agate are less smooth.

While these silica minerals are the best known and most often found in collections, they are not the most common form of this very common substance. The transparent, colorless variety of opal called hyalite occurs in small seams and fractures of innumerable rocks where it precipitated from mineral-laden ground water, picking up a trace of a uranium salt as it did so. A few molecules of uranium per million of silica are sufficient to activate fluorescence. The glow is not evenly distributed but varies from dull deep green to bright yellow-green. Probably more rocks show traces of this familiar color than any other.

There may be some areas where hyalite is not fluorescent, but it is never markedly phosphorescent. Usually it is not smooth to the touch, since its thin film closely follows the texture of the rock. Sometimes hyalite may hide the fluorescence of underlying minerals by its own strong glow. Always check a strange fluorescence to make sure beauty is not merely skin-deep.

Later we come upon a gorgeous sight—a narrow ledge that lights up with a deep orange radiance. The sunset color comes from rock that seems to be crumbling away. Extracting a piece the size of your fist you notice its fluted appearance; long slender crystals or columns give the appearance of a bunch of matches tightly bound together.

A quick "charging" with the ultraviolet lamp at point-blank range leaves an afterglow as the lamp is withdrawn. The fading phosphorescence is white, not orange. A drop of acid would fizz like soda water on contact with it, proving it a carbonate like calcite. Actually it is aragonite, differing from calcite only in form but sharing the same wide-traveling habits.

Aragonite has a more interesting structure and fluorescence than most calcites. Its color under the lamp may range from pink or cream to deep red, sometimes interspersed with stripes of yellow or orange. Green fluorescence is seldom found, but the phosphorescence sometimes has a greenish-white cast. Where aragonite's phosphorescence is the same color as the fluorescence, it is usually less brilliant and fades quite fast, especially the colors at the red, pink, and orange end of the spectrum.

Our specimen, though somewhat crumbly, is handsome, glowing in tones ranging from deep orange to pale rosy pink. Wrap it in your handkerchief to take along.

You almost stumble over a hard round object half-buried in the sand. The flashlight shows a rough, warty, potatolike ball. You crack it a sharp blow with a hammer. Splitting in half, it reveals itself as a hollow shell lined with innumerable sharp crystal points flashing in the light. The lamp's beam instantly works a startling change: the inside of the shell glows fiery red, almost as if red-hot, yet it is perfectly cold to the touch. You remove the lamp, but the illusion of heat persists until the phosphorescent afterglow has faded.

This is a geode, a gas pocket in an ancient volcanic deposit, later penetrated by percolating waters that deposited their dissolved mineral burden inside as a fluorescent lining. You have found a handsome specimen, especially so since the fluorescent interior of a geode usually is one of the silica minerals, seldom showing fluorescent response other than uranium green. This bright-red crystalline interior is rare indeed; it will prove a treasure to possess as long as you live. Lady Luck has really been with us tonight.

Fluorescent Locations are Everywhere

It is getting late, so we decide to resume our explorations of the quarry on some other visit. If time permitted, our imaginary trip might well take us to some of the prime hunting grounds where some of the finest fluorescent specimens have been found. We might travel to the vast volcanic plains of Oregon, Montana, and Wyoming, where ancient ground waters filled innumerable "thunder egg" nodules with agate; or to Nevada's Virgin Valley, where a far-spread layer of sea-green glowing opalite *(plates 12, 13)*, topped by a crust of white volcanic ash, looks like an enormous lime-gelatin salad.

Or we might visit Arizona's mining areas, where a collector's favorite reward is a water-clear quartz crystal completely enclosing a tiny, perfect crystal of scheelite that lights up with splendid blue inner radiance: a jewel within a jewel; or we might go to a famous limestone quarry near Clay Center, Ohio, where pockets are filled with a rich creamy-fluorescing fluorite associated with fragile blue-and-white blades of celestite.

We might go to a pegmatite near Grafton, New Hampshire, where a giant halo surrounding a primary uranium deposit offers fernlike growths of brilliant green-fluorescing autunite against a background richly colorful with radioactive minerals in brown, orange, yellow, and green.

But each mineral hunter has the opportunity to blaze new trails. In rich farmland areas made up of the sediments from ancient rivers, alert collectors find hunting grounds in clay banks and brickyards, or along lake shores and stream beds. Water-well digging brings up a great variety of minerals and sometimes fossil shells that may be fluorescent.

Even city dwellers in the East, who rarely see mountain or desert, carry on their prospecting in limestone quarries, gravel pits, and railway cuts. Slag piles from foundries often provide collector's treasure. So do the yards where ornamental rock is sold for gardens or as cut stone. An eastern mineral society alerted its membership to a forthcoming tunneling project whose mile-long burrow would cut many formations and yield waste piles for investigation.

Not all the rich fluorescent areas have been found, experience shows, for new localities still are being discovered. More will be located by alert collectors. Not long ago, several gorgeous new fluorescent minerals were reported in Texas and California in deposits around hot springs and fumaroles, the fluorescence being caused by hydrocarbons brought up in the hot vapors from below *(plate 16)*.

The beautiful eucryptite, whose delicate rose-pink fluorescence was discovered in South Africa, has just been reported in Wyoming. Mineralogists say this rare mineral is far more common than was formerly believed. Throughout the country, many well-trodden prospecting grounds have been inadequately explored with the ultraviolet lamp.

By all means, widen your knowledge by joining a mineral club. Most of them go on regular field trips to localities known to be promising. You will learn much and will find new friends with common interests.

Take the whole family along; mineral collectors are friendly people and interest in the gorgeous minerals is contagious. You will find that the out-of-doors activity and the satisfaction of exploring new fields of learning provide a welcome relief from the tensions of the workaday world.

Fluorescence of Ore Minerals

A beam of ultraviolet light, invisible in the darkness,

strikes a drab, nondescript rock. Instantly, patches of gleaming deep yellow freckle the dull surface like flakes of gold. This is the typical fluorescent color of zircon.

Now a whitish rock resembling ordinary limestone is brought into the beam. Tiny blue-white specks glow against the deep purple background like stars in a midnight sky. They announce the presence of scheelite, ore of tungsten *(plate 7)*.

Next, a white chalky mass placed under the lamp likewise turns brilliant blue. So do one's fingertips, where the soft mineral has rubbed off on them. But pure scheelite is hard and does not rub off. This must be hydrozincite *(plate 9)*, an important zinc mineral.

Such is the magic of ultraviolet light. When the mercury-vapor lamp, a rich source of potent "black light", was provided with a filter to screen out unwanted visible rays and adapted to battery-powered, lightweight, portable outfits, it opened up an amazing new world to collectors and prospectors. Soon many pioneering mineral hunters were in the field, anxious to try out this new instrument in locating ore.

This was in the late thirties. The timing could not have been more fortunate. On a fateful day in December 1941, United States military leaders suddenly faced the fact that the lifeline leading to China, the nation's principal source of tungsten, was broken. Without this hard, heavy metal to toughen big gun barrels and armor plate, no nation could adequately equip a modern army and navy.

Tungsten ores always had been hard to identify. Fortunately, prospectors had discovered that with their mineral lamps, detecting calcium tungstate—scheelite—was merely a matter of pushing a button. No one knew if the United States had any substantial deposits of scheelite, but bonuses for new discoveries quickly brought ore seekers into the field by hundreds. Many new strikes were made. Old mines were re-examined, yielding new treasures.

Geologists of the U.S. Bureau of Mines, seeking new sources of antimony, started test-drilling on an Idaho property owned by the Bradley Mining Company. As routine, all cores were examined under ultraviolet light. Soon appeared a series of cores heavily sprinkled with brilliant blue-white specks. An enormous orebody had been tapped—not of antimony, but tungsten. So near the surface it lay that steam shovels were moved in for open-pit operation. Thus the famous Yellow Pine Mine operated night and day throughout World War II, setting a nationwide record for tungsten production.

This was but one of hundreds of strikes that broke the tungsten bottleneck and added undreamed-of reserves to the national supply.

Uranium—Another Target

History repeated itself when the great postwar uranium rush made the little lamp with the dim purplish beam a working partner of the Geiger counter. Experienced miners saw new possibilities. With it they searched formations for the characteristic yellowish-green glow that instantly pinpoints streaks of fluorescent minerals in "hot" formations. They used it for quick fluorescent bead tests to distinguish between uranium and thorium. It could not be contaminated by uranium dust in mine workings or damaged by seeping radon gas. It guided them in placing charges when blasting, kept them on the vein when excavating, and was invaluable in sorting high-grade ore from leaner stuff.

The inexperienced found that the probing beam of the ultraviolet lamp could make many ore minerals stand out distinctly from apparently barren rock and from many of the bright-colored but worthless minerals found in uranium country. It enabled them also to recognize nonradioactive minerals that are hard to identify without laboratory tests. Moreover, it did not give false readings on deposits containing radioactive phosphates, zircon, or thorium minerals.

Soon they were picking up traces of other ores. A number of them struck unexpected deposits of tungsten ore and promptly began harvesting this valuable money crop. Even a small vein that two men could work together by hand could provide an ever-available bank of funds to grubstake further prospecting.

News of strikes of zirconium ore, fluorspar, and ores of lead, zinc, beryllium, niobium, tantalum, and the newly important rare earths began to hit the headlines as by-products of uranium hunts.

Finding Overlooked Treasures

Growing interest in ultraviolet prospecting thus brought a wave of new discoveries in old mining districts. One ghost-town area was reborn when a Nevada rancher borrowed a mineral lamp from a friend and set out to look for the blue-white glow of scheelite. His first sight was a glowing yellowish-green mass. Fortunately, he took back samples of the pretty rocks. When friends identified them as autunite *(plate 8)* and the Geiger counter went wild, a new uranium rush was born.

An Arizona county assessor who collected rocks as a hobby bought an ultraviolet lamp mainly out of curiosity. Immediately he faced a maddening problem. Among his specimens was a large chunk of white quartz, splotched with areas that under the lamp shone clear blue as the sky at dawn.

Where, in the years of random tramping over the desert, had he picked up this rich piece of tungsten ore?

For months the question haunted him. He revisited all the places he could remember. Finally he found the source in an abandoned gold mine. The overlooked pay rock not only lined tunnel walls, it outcropped in untapped deposits and lay in heaps on the dump, already mined.

An engineer, chatting with the owner of a Utah mine, remarked “this ore looks as if it might contain scheelite.” The owner, having heard about ultraviolet prospecting but nothing about how it could apply to a property yielding only gold, silver, and copper, nevertheless bought a lamp and walked through the workings. On every side the wall rock blazed with blue fire. It seemed completely lined with scheelite.

A major company’s chief geologist investigated and returned breathless. “Never saw anything like it—practically a solid mountain of tungsten!”

It was one of the biggest strikes in years.

New uses for ultraviolet lamps continued to multiply. Experienced prospectors take advantage of the principle that “minerals of a feather flock together.” Lead and zinc ores, for instance, traditionally go together like ham and eggs. Thus the prospector whose ultraviolet lamp strikes up the pale-yellow glow of anglesite in an outcrop instantly knows chances are good of finding (1) other members of the lead family—the important cerussite, which sometimes fluoresces a deeper yellow color, or galena, which does not fluoresce at all—and (2) zinc minerals as revealed perhaps by the rare green glint of willemite, the uncommon creamy yellow fluorescence of calamine, the bright blue of hydrozincite, or the occasional deep orange radiance of sphalerite.

By similar reasoning the prospector can read the signs that a formation lies in the oxidized zone of a vein deposit. The fluorescent colors of leadhillite, smithsonite, or autunite thus may point the way to underlying veins of galena, sphalerite, or uraninite, characteristically belonging to the deeper, unoxidized zones.

Lost veins thus become easier to locate, for unimportant

minerals in oreshoots can serve as fluorescent tracers. Common calcite, for instance, may lend its ready fluorescence to outline the ore-bearing vein as a glowing white or pinkish stripe against common country rock. Fluorite, dolomite, and barite may offer glowing threads of color to trace the path of the mineralized zone.

New Fun for the Collector

The nation's amateur mineralogy fans already had claimed the lamp for their own. Its theatrical effects in a darkened room dramatized their mineral treasures. It converted dull rocks into miracles of glowing color to fascinate even the most indifferent spectator. Amateur collectors formed the habit of putting the lamp on every interesting rock they encountered while specimen hunting. In so doing, they were rewarded by many unsuspected finds and demonstrated that an unknown mineral that glows under ultraviolet light has already partially or wholly identified itself.

More will be said about these techniques in the pages that follow. First, however, for the benefit of the newcomer to ultraviolet light and its marvels, a brief initiation is in order.

Fluorescence at Night is a New Experience

It is always best to do your mineral hunting at night. The eyes become dark-adapted and can pick up weak fluorescence at greater distances. Even under close examination with the lamp, more distinguishing factors can be observed.

Night field trips have a freedom and excitement all their own. You must walk cautiously and avoid planting your feet on a crumbling trail edge or backing into unnoticed cactus; but even on a clear moonlit night, most fluorescent minerals should be recognizable within a 3-foot circle of your medium-powered shortwave lamp. More sensitive minerals, such as autunite, scheelite, hyalite opal, and hydrozincite, may show up as far away as 20 feet. On an inky black night, they sometimes can be seen gleaming at 50 to 75 feet.

By night you will discover the meaning of "invisible" fluorescent minerals. In daylight an apparently ordinary rock specimen may show several types of minerals of little or no interest, but in the dark, under strong ultraviolet light, certain unnoticed specks or transparent crystals become dominant. They seem almost to start from the rock in their strong contrast to the dark matrix.

Switch on the white light and they are gone, for their color blends in so closely with the adjoining rock that they are lost to sight.

Examine a handful of sand for an especially clear demonstration. Under white light, the reflection from thousands of shiny grains merges into a bright mass in which no single crystal can be distinguished. Now turn off the light; switch on the ultraviolet lamp. One lone grain glows brilliantly among its dark neighbors. The single crystal has become a powerful primary light source. The effect is the same as that of a lighted match in a desert wilderness on a starless night. It can be seen for great distances.

These tiny fluorescent beacons appear to give off a great deal of fluorescent energy, but actually they are not strong by our normal standards of white visible light. The eye, dark-adapted, suddenly finds its powers vastly expanded and vision is many, many times more acute. This is why the ideal time to do your fluorescent-mineral hunting is by night.

The ultraviolet lamp reveals fluorescent minerals in sharp detail with startling clarity, but take care to keep it as close to the rock as possible. Don't work too fast or you will miss faint showings. When in doubt, take a sample and examine it in total darkness, allowing time for your eyes to adjust themselves. On a bright day your pupils contract to pinpoints but in darkness they will soon open wide.

In daylight you probably will sample promising outcrops as you go, perhaps checking from time to time with the lamp in the darkness of a viewing box, or under a coat or large black cloth. Or you may defer lamping until you return to camp, in which case you will want to tag samples with marked adhesive tape or drive a stake to mark the location. Scrawl trail directions on the rocks with fluorescent chalk to guide you when you return by night for intensive lamping.

The Choice of a Lamp

Fluorescent minerals fall naturally into two groups: those which fluoresce under longwave ultraviolet rays of the 300-400-nm range, and those that respond only to shortwave ultraviolet light, ranging downward from 300 nm.

An interesting assortment of minerals responds chiefly to the longwave lamp. It produces softer pastel colors, but you will find that many fluorescent minerals do not respond to it at all.

Minerals that fluoresce under the shortwave lamp are much more numerous than those that respond to the longwave lamp. The shortwave lamp brings out richer, stronger, more contrasting colors. At least some response is noted with nearly all fluorescent minerals. Most commercial fluorescent ores are brightest under the shortwave lamp, whose stronger effects speed fieldwork and permit prospecting at greater range. For fieldwork, if a choice must be made, select the shortwave variety.

Alternate use of longwave and shortwave lamps can give valuable help in home or laboratory mineral identification. For instance, the green glow of hyalite opal is often mistaken for the uranium mineral, autunite, but hyalite seldom fluoresces under longwave ultraviolet light. Autunite answers to either.

Interesting display effects can be exploited with certain minerals that not only have *two* fluorescent colors but also show afterglow in a third color. For example, a calcite from Miami, Florida, fluoresces and phosphoresces pale green in shortwave ultraviolet light, but under longwave ultraviolet light it glows bright yellow-white, followed by phosphorescence.

Certain remarkable calcites found at Terlingua, Texas *(plates 2-4)*, give a momentary pink fluorescence when the shortwave lamp is snapped on, but this color quickly disappears. It is swallowed up by a blue fluorescence that seems to build up until it is far brighter than the pink glow. With a strong lamp, this blue glow continues as a phosphorescent afterglow, persisting as much as a minute, so bright one can read newspaper headlines by it. Switching to the longwave lamp produces an immediate pink fluorescence but no afterglow.

True Fluorescence or Reflection?

A great many minerals, especially white ones and transparent glassy crystals, appear somewhat purplish under the ultraviolet lamp. A few minerals, particularly dumortierite, sometimes show a true fluorescence of a purplish to violet hue. Usually, however, such a color is merely a reflection of a small amount of purple light that passes through the filter. To make sure, place a piece of window glass between the lamp (shortwave only) and the rock. If the glowing color persists, it is merely a reflection of visible light, not true fluorescence, since shortwave ultraviolet light does not go through the glass. Any report of a "purple fluorescence" is justifiably open to suspicion.

So also is the "black fluorescence" reported by some early investigators. Black is not essentially a color, but merely the absence of light, so that any such expression is meaningless. Such specimens simply are absorbing *all* light falling upon them, hence they look black.

Novices sometimes are surprised, when lamping their specimens after arriving home, to discover glowing purplish spots they had not noticed before. These are rock "bruises." They are one reason the experienced collector always takes along material to wrap choice pieces or otherwise prevent them from knocking together in transit.

More than one prospector has used discarded rocks to enclose a campfire, later discovering that the rocks have become strikingly fluorescent. This sort of magic occurs especially with limestone, which becomes thus altered by the heat of the fire. The dumps of old mining camps often show fluorescent smelter slag.

In lamping, remember this:

1. Give your eyes time to adjust themselves to the dark.
2. When testing for phosphorescence, close your eyes as you count to at least ten—or still better, twenty. As you charge the mineral, you also recharge your eyes for greater efficiency. By removing the filter from the lamp you can give the rock a still stronger charge.
3. If doubtful about a color observed under the shortwave lamp, see if putting a piece of window glass between lamp and specimen stops the glow.

Fluorescent Gemstones and Lapidary Material

The world of gems takes on a new and unexpected appearance when explored under ultraviolet light.

Blue sapphires may glow red; pink tourmaline turns blue. Priceless gems may appear pale and lusterless, while cheap imitations shine with searchlight brilliance.

Why this strange, capricious behavior? Why do rubies from Burma glow such a strong red color, while those from Siam usually show a weaker fluorescence? Why do diamonds from many lands *(plate 1)* fluoresce brilliant blue-white, others fail to glow at all, and those from a few areas show hues ranging from reddish-orange to pale green? The answers to some of these questions may be found in the chapter "Activators in Fluorescent Minerals", following in this book. For now, though, let's turn our attention to some other aspects of fluorescent gems.

Gem Analysis and Grading

Gem collectors find ultraviolet light helpful in examining and purchasing stones. A stolen gem can be recut to disguise its identity, but its fluorescence is not altered and may betray its locality of origin. Certain temperamental variations in fluorescence, often found in natural gems, can help distinguish them from synthetic stones. For instance, a yellow sapphire that fluoresces is certainly not from Australia and probably not synthetic.

Imitation gems such as garnet-and-glass doublets quickly reveal themselves as such under the ultraviolet lamp. The glass base usually fluoresces bright yellow-green or apple-green in strong contrast to the nonfluorescent garnet top. Cement sticking together the components of doublets and triplets usually gives a telltale glow.

Other materials used in imitation gems include celluloid, which fluoresces yellow under both the longwave and shortwave lamps, and casein, which fluoresces blue to white under both lamps. Bakelite, much used in imitation amber, is nonfluorescent.

For grading of precious stones, fluorescence is not a primary tool, since commercial value rests chiefly on other factors. The lamp can, however, sometimes expose flaws or cleavage planes in certain stones. Zones or planes of glowing color may reveal unsuspected impurities introduced during growth, as well as inclusions whose presence may have a bearing on the identity of the gem.

Fluorescent gemstones can be viewed best against a dark, dull, nonreflecting surface such as black velvet.

Fluorescent Color—
A New Dimension in Gem Cutting

If beauty, rather than facets, is the prime attribute of a gem, surely many of the gorgeous fluorescent stones merit this rating. Few jewels can outshine the powerful blue blaze from a clear natural crystal of scheelite or the brilliant yellow-green glow of clear, colorless opal under the shortwave lamp. Extra drama comes from the fact that in daylight they look very much like glass, but suddenly spring to life in the dark when touched by ultraviolet rays.

Faceting has its limitations, for the brightest of glowing minerals are usually softer than quartz and do not stand up too well in the faceting operation. Some are very susceptible to abrasion, or they cleave too readily to withstand shocks. But for display purposes, they add great distinction and rare beauty to a collection.

Almost any fluorescent mineral's brilliance can be enhanced by polishing its best surface. By thus smoothing out shadows, brightness can be raised by 25% or more.

A skillful cutter can achieve striking success in faceting many minerals not usually recognized as gems. The U.S. National Museum in Washington, D.C., contains a showcase of faceted stones, among which are axinite, apatite, willemite, sphalerite, danburite, vesuvianite, petalite, orthoclase, scheelite, diopside, and kyanite. Another display of rough and cut stones shows cabochons of aragonite from California, green diopside from New York, thomsonite from Minnesota and thulite from Norway, as well as yellow and blue smithsonite.

All these minerals have been found to fluoresce in certain localities. This little-explored field offers challenging opportunities to the craftsman in search of fluorescent specimens with the makings of a fine cut stone.

Massive cuproscheelite, fluorite, and aragonite have excellent possibilities for gem cutting. The bright yellow fluorescence of wernerite is shown to great advantage in cabochons. Some uranium-bearing quartz specimens hold hidden fluorescent fire. Others have interesting fluorescent inclusions or show phantoms and colored-zone effects under the lamp.

Intergrowths of two or more fluorescent minerals make especially fascinating cabochons *(plate 31)*. Sphalerite intergrown with wurtzite, as found at Frisco, Utah, is a dramatic example. Some of the finest combinations come from Franklin, New Jersey: such groupings as deep-red calcite, strong yellow esperite, and brilliant green willemite. Franklin willemite is an outstanding favorite for its penetrating green radiance and a strong phosphorescence that sometimes persists for hours.

Cutting of Fluorescent Gemstones

Certain precautions should be observed in cutting fluorescent minerals. Kerosene or other cutting oil must be removed before one can judge the results of one's work, since most petroleum derivatives fluoresce, usually blue. Given several days, kerosene will evaporate; or it can be washed off with soap and thoroughly rinsed. Trichloroethylene is an excellent solvent for oils.

Before washing, be sure to examine the cleansing agent itself with the lamp before using it. Many widely known detergents fluoresce strongly.

An interesting discovery by Hugh Leiper, F.G.A., of Austin, Texas, illustrates one use of fluorescence in judging cutting material. Leiper, in faceting much blue topaz while building up a collection representing every known country of occurrence, was often baffled by failure to get a satisfactory flooding of the gem with the beautiful sky blue for which the material is known. Sometimes the finished gem would show this color only when tilted to certain angles away from the perpendicular line of sight.

One night, passing his shortwave lamp over a tray of large topaz crystals, he noticed a lemon-yellow fluorescence from a large cleavage of Australian topaz. On closer examination the glow seemed to come from an hourglass-shaped area extending from two opposite corners of what originally was an almost square crystal.

Going over his collection of rough gem topaz, Leiper discovered that many crystals fluoresced in one of three distinct patterns, giving an excellent standard for grading topaz in the rough. Oriented with proper reference to the fluorescent zones, facets could be cut so as to obtain the finest depth of color obtainable.

Leiper advocates working with the rough stones in complete darkness, having first made sure they are free of oil or detergents, and then examining them under shortwave ultraviolet light to orient the color zones accurately in the preform. Thus the finished gem will display its color to maximum advantage.

Unusual Lapidary Projects

Sphere-cutting enthusiasts will find that a perfect job becomes even more noteworthy if the cutting material is fluorescent. A group of fluorescent spheres under strong ultraviolet light becomes a glowing constellation of bright colors, with bandings giving added emphasis. Unusual possibilities lie in spheres cut from the brilliant Franklin minerals. The famous red-green-black combination of calcite with massive willemite and franklinite is spectacular, but can be made really breathtaking by mounting such a sphere on a spindle so it can be spun rapidly, merging the fiery colors into ever-changing whorls of color.

If you have slabbing equipment, you will find fluorescent book ends a really attractive oddity. Some ambitious lapidaries have produced tabletops that become showpieces when an overhead ultraviolet lamp brings the mosaic-like pattern of inlaid rock chips to life with fluorescent colors. The material can be imbedded in plaster or cement that after setting is ground down to smoothness.

A stairway landing or dark patio floor can be made safer and highly glamorous if topped with plaques or panels of terrazzo in which fluorescent calcite, wernerite, or other fluorescent mineral fragments are imbedded. A concealed ultraviolet lamp just above floor level provides the needed invisible rays.

Some rockhounds have built fireplaces whose outer surface is faced with sections containing slabbed fluorescent rocks. If a rustic effect is desired, only the back side of each rock is cut level, leaving the best mineralized face exposed.

A California builder of swimming pools provides a fluorescent coping surrounding each pool. Lighted with double 40-watt tubular ultraviolet lamps, it gives a gorgeous border that at night offers an additional factor of safety.

Your artistic ability can find an interesting outlet in a new medium for picture making, by using fluorescent grains or chips. These are laid down on a surface that is thinly coated with an adhesive, such as model airplane cement or water glass (sodium silicate). For small pictures, sandlike grains can be used, but better effects are achieved by use of chips applied to a larger canvas *(plate 15)*.

The Franklin Minerals

For the collector of fluorescent minerals, no locality holds the same magic as the fabulous zinc mines of Franklin, New Jersey.

Franklin has yielded more different fluorescent minerals than any other locality, some of them exclusive to this locality. In beauty and brilliance, beyond doubt, they are unsurpassed anywhere. In complex mineralization and kaleidoscopic changes that took place during its formation, Franklin is like no other ore deposit.

As the mineralogical richness of the remarkable deposit became known, famous scientists from all over the world flocked to study its wonders; yet it must be said that even today much of its geological history is yet to be fully explored and interpreted.

Long before mineral collectors had become aware of the beauties of fluorescence, countless tons of "waste" rock—including the spectacular red-fluorescing gangue material (massive calcite) and many rare associated minerals—went into the paving of roads in the New Jersey countryside. Even today some roads are well ballasted with specimen material.

To the anguish of hungry modern collectors, some of the finest treasures probably still lie abandoned in the workings, which have been closed and surrendered to the creeping waters. A new mine[1] now is operating three miles away and a modern mill has been built alongside the shaft, but improved methods have done away with the famous picking tables. Ore

---

[1] The Sterling mine in Ogdensburg, New Jersey, closed in 1986, and by late 1993 most of the underground workings were flooded. However, the Sterling Hill Mining Museum now occupies the same site. Visitors may take a tour of the surface level of the mine, view fluorescent minerals still in place underground, and examine many excellent displays in the main museum building.

is crushed underground, and the material brought up from below yields only a floury white sand as waste material. The dumps that have been combed by thousands of rockhounds down through the years have virtually disappeared. Only the minerals gathered in years gone by *(plates 17-31)* are available to the new generation of fluorescent-mineral enthusiasts.

Photographs of Fluorescent Minerals from Worldwide Localities

Plates 1-16

Plate 1

Plate 1. Diamonds from worldwide localities, showing the typical range in fluorescence color for the species: bright pale to medium blue, yellow to green, and more rarely orange to red. The photograph was taken under longwave ultraviolet light. Most of the diamond grains are 2-4 mm in diameter. Photo by David Grigsby, courtesy of the Gemological Institute of America.

Plates 2-4. A mass of granular calcite about 6 inches long from the Terlingua mining district, Brewster County, Texas. Though of undistinguished appearance in daylight (plate 2), Terlingua calcite is deservedly famous for its remarkable fluorescence, which changes from bright medium blue under the shortwave lamp (plate 3) to bright medium pink under the longwave lamp (plate 4). Equally remarkable is the strong and persistent blue phosphorescence following shortwave excitation. Specimens when first collected are white to pale green in daylight, but the green areas upon exposure to sunlight gradually turn pink, as shown in plate 2. The fluorescence seems not to be affected. David Grigsby photos.

Plate 2

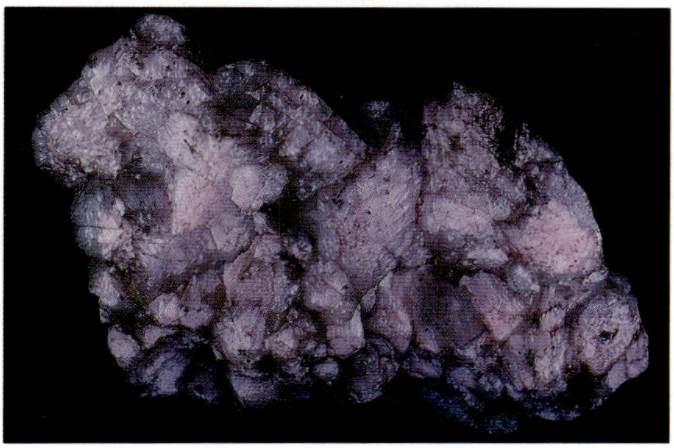

Plate 3

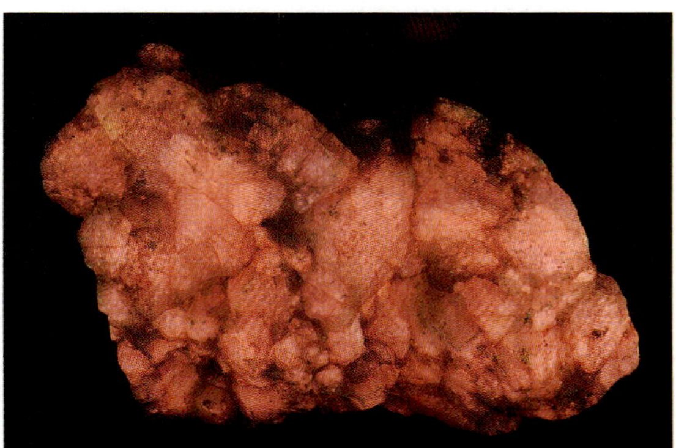

Plate 4

Plate 5

Plate 5. Massive, granular colemanite from the Calico Mountains, San Bernardino County, California, showing the typical white to orange-yellow fluorescence of this mineral under longwave ultraviolet light. The specimen is about four inches across.

Plate 6. Tugtupite from the Ilímaussaq intrusion, Kvanefjeld, Greenland, showing the fine, deep-red fluorescence for which the species is so well known. This specimen, an uncommonly rich example, is about two inches across and is shown here under shortwave ultraviolet light. Under longwave ultraviolet light the fluorescence is equally bright but of much different color: salmon orange. Tugtupite is also one of the relatively few minerals that exhibits pronounced reversible photosensitivity, or tenebrescence: the pink to raspberry-red daylight color deepens during exposure to shortwave ultraviolet light and then fades as the specimen is kept in darkness, or exposed to incandescent light.

Plate 7. Rich scheelite ore from a tungsten mine near Hawthorne, Mineral County, Nevada. The polygonal outlines of some of the scheelite grains are sections through sharply formed crystals embedded in the rock. The specimen was photographed under shortwave ultraviolet light and is about four inches across.

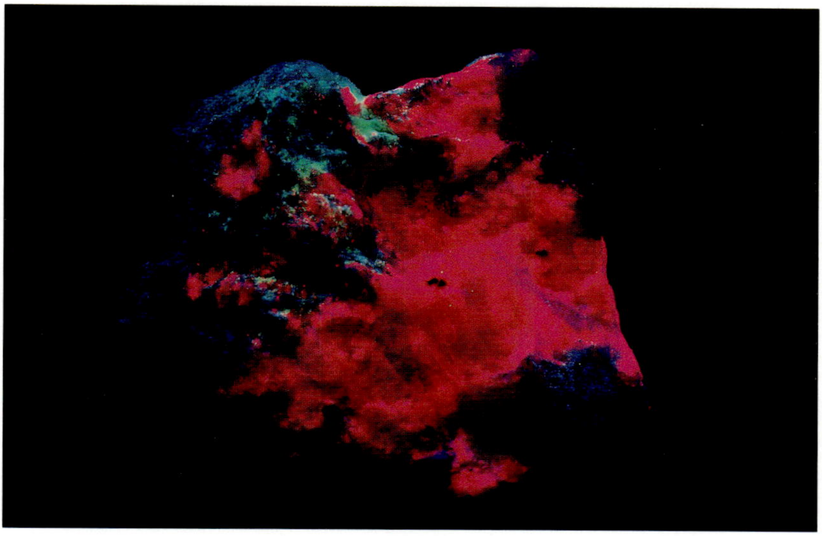

Plate 6

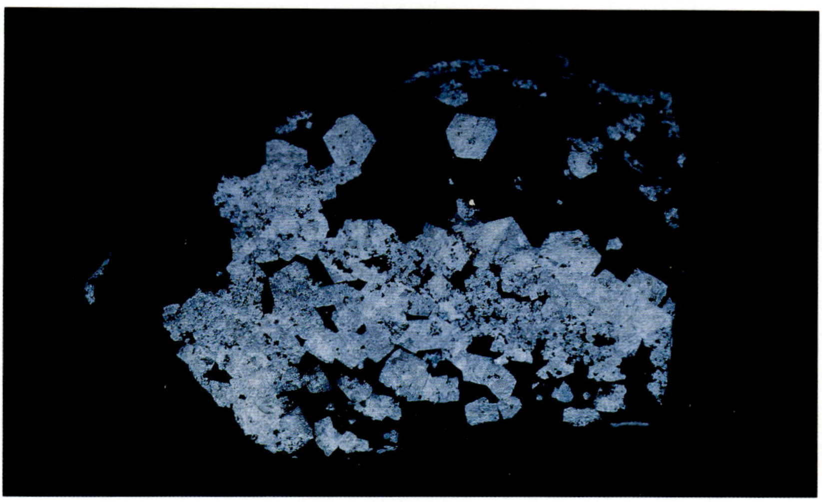

Plate 7

Plate 8

Plate 8. A fine specimen of intergrown, platy crystals of meta-autunite from the Daybreak mine on Mt. Spokane, Spokane County, Washington. Meta-autunite under shortwave ultraviolet light is among the brightest of all known fluorescent minerals—so bright, in fact, that its fluorescence sometimes overwhelms that of other specimens in a display case. The Daybreak mine is the premiere U.S. locality for the species and has produced magnificent plates and slabs, some more than two feet across, of sharply formed crystals. This specimen is about three inches long. David Grigsby photo.

Plate 9. Hydrozincite, in tiny white crystals lining vugs, from the Goodsprings mining district, Clark County, Nevada. The specimen is about four inches across and was photographed under shortwave ultraviolet light. David Grigsby photo.

Plate 10. Granular sodalite from Mont St. Hilaire, Quebec, Canada. The lovely orange fluorescence of sodalite under the longwave lamp is matched by that of few other mineral species. This specimen, a sawn slab about six inches across, is a particularly rich and attractive example. David Grigsby photo.

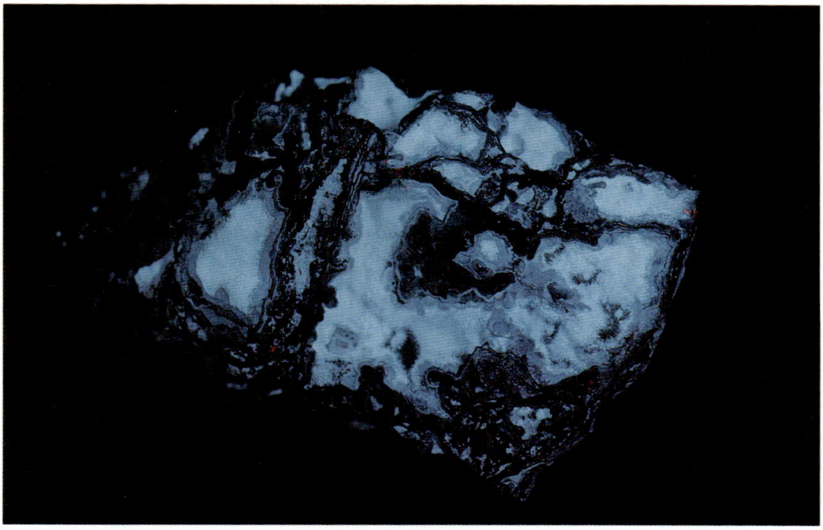

Plate 9

Plate 10

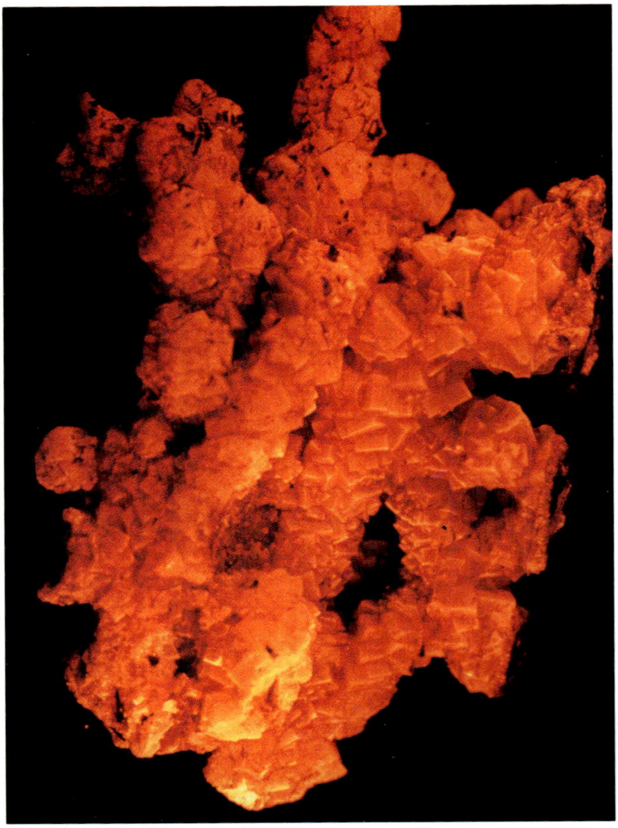

Plate 11

Plate 11. Halite, in intergrown cubic crystals encrusting sagebrush twigs, from the Salton Sea, Imperial and Riverside Counties, California. The bright orange-red fluorescence of this unusual material is activated by small amounts of manganese, with lead as coactivator. The specimen, photographed under shortwave ultraviolet light, is about ten inches high.

Plates 12-13. A finely layered mass of opal (variety common opal) from Virgin Valley, near Denio, Humboldt County, Nevada. Virgin Valley opal formed as a replacement product of extensive beds of volcanic ash and is a popular collector's item; the bright yellow-green fluorescence is due to minor amounts of uranyl ion as an impurity. This specimen, pale brown in daylight (top), is about four inches thick and was photographed under shortwave ultraviolet light (bottom). The fluorescence is nearly as bright under the longwave lamp.

Plate 12

Plate 13

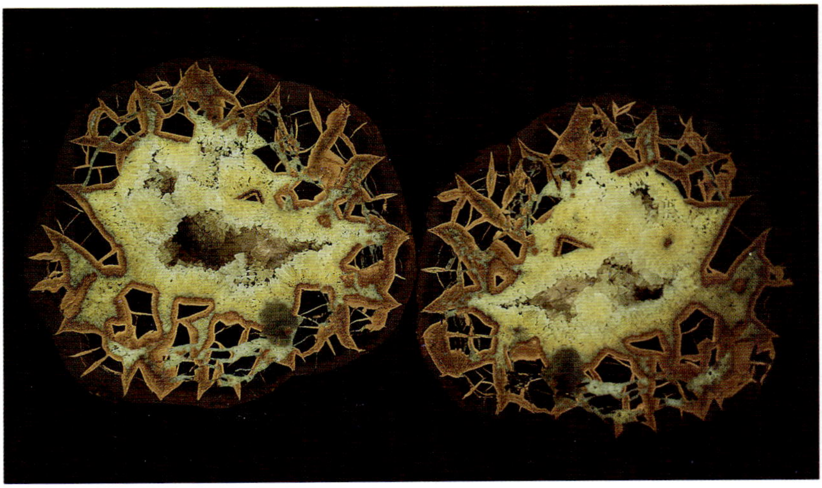

Plate 14

Plate 14. Two halves of a sawn and polished septarian nodule from the diggings about five miles south of Orderville, Kane County, Utah. The yellow-fluorescing crystals lining the interior are calcite; the orange- to brown-fluorescing mineral filling cracks and rimming the broken rock fragments reportedly is aragonite. Many thousands of nodules have been produced from this deposit. This fine example, about eight inches in diameter, was photographed under longwave ultraviolet light.

Plate 15. Fluorescent minerals in art: a lovely group of flowers made of granulated fluorescent minerals viewed under shortwave ultraviolet light. The painting measures 9.5 x 7.5 inches. David Grigsby photo.

Plate 16. Small, radiating, bladed crystals of karpatite from the Fourth of July mine, New Idria mining district, San Benito County, California. Karpatite, a rare hydrocarbon mineral, is much prized for its brilliant blue fluorescence. This specimen, a particularly attractive example, measures about two inches across and was photographed under shortwave ultraviolet light. Henry Van Lenten photo.

Plate 15

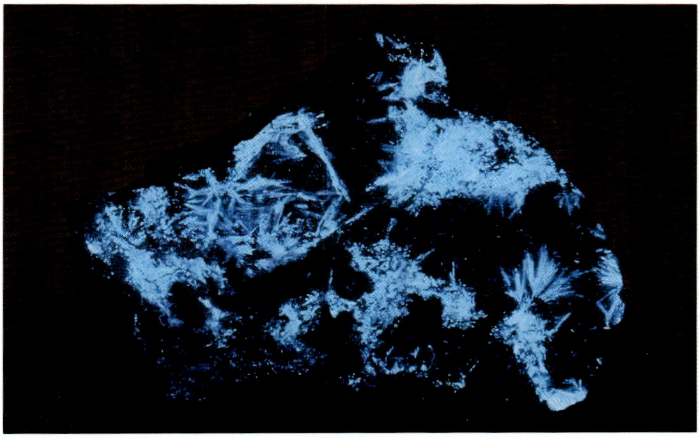

Plate 16

Photographs of Fluorescent Minerals from the Franklin-Sterling Hill District, New Jersey

Plates 17 - 31

Plate 17

Plate 17. A good example of a "four-color" specimen from Franklin: hardystonite (fluorescent deep violet-blue), clinohedrite (orange), calcite (red), and willemite (green). The specimen, about 4.5 inches across, is shown under shortwave ultraviolet light. Henry Van Lenten photo.

Plate 18. Radiating willemite in a dolomite-serpentine vein in compact franklinite-willemite ore from Franklin. Specimens of secondary willemite such as this one are highly prized for their astonishingly bright and persistent phosphorescence, which commonly remains visible for hours after exposure to the shortwave lamp. This specimen, photographed under shortwave ultraviolet light, is one of the finest known and measures more than 7.5 inches across. Henry Van Lenten photo.

Plate 19. An irregular mass of turneaureite (fluorescent orange) embedded in calcite (red). Turneaureite is among the rarest of the brightly fluorescent minerals from Franklin. Rich specimens of large size are few; this one, shown under shortwave ultraviolet light, is ten inches long. Henry Van Lenten photo.

Plate 18

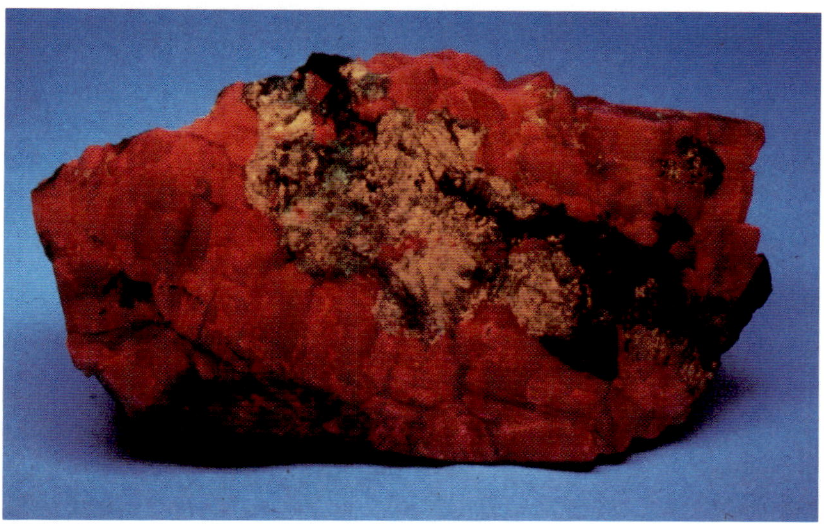

Plate 19

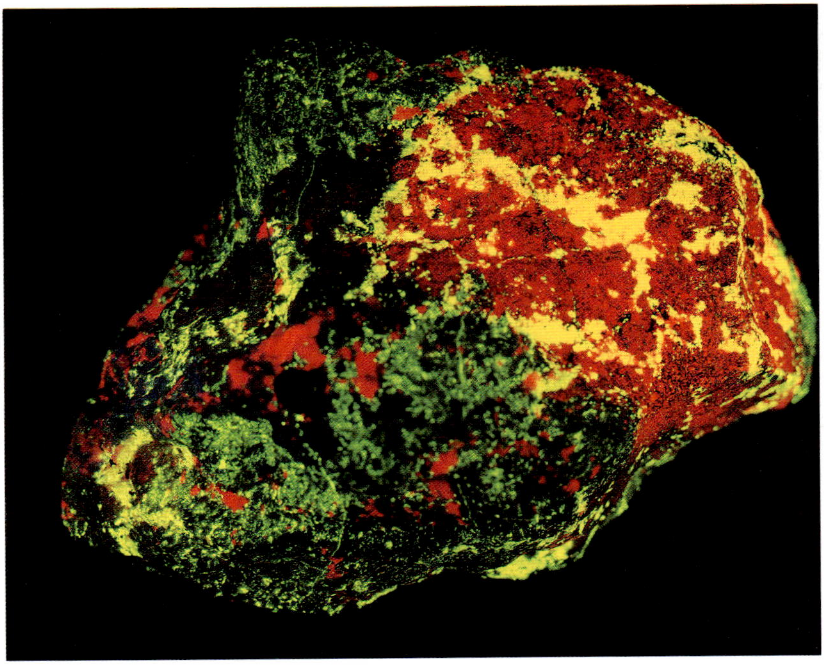

Plate 20

Plate 20. Esperite (fluorescent yellow), calcite (red), willemite (yellowish green), and minor hardystonite (deep violet-blue) from Franklin. Specimens such as this five-inch example, shown here under the shortwave lamp, are prized for their bold, contrasting colors and textural complexity.

Plate 21. Granular, white-fluorescent barite enclosed in red-fluorescent calcite from Franklin. The barite layer is about one inch thick and was photographed under shortwave ultraviolet light. Specimens this large (5.2 inches tall) and rich in barite are true rarities, much coveted by collectors. Henry Van Lenten photo.

Plate 22. A rich mass of granular, white-fluorescent barite in red-fluorescent calcite from Franklin. This is a fine example, notable for the large size, abundance, and bright fluorescence of its barite grains. The specimen measures about five inches across and is shown here under shortwave ultraviolet light. Mitch Dolobowsky photo.

Plate 21

Plate 22

Plate 23

Plate 23. Massive, coarse-grained sphalerite showing a fine, pinkish-orange fluorescence with blue highlights under longwave ultraviolet radiation, from 10 pillar below 2350 level, North Ore Body, Sterling Hill. The specimen is about three inches high. Henry Van Lenten photo.

Plate 24. Margarosanite (fluorescent medium blue through bluish pink) with willemite (yellowish green) and minor clinohedrite (orange-fluorescing mineral bordering margarosanite toward base of specimen), from Franklin. This specimen, about five inches across and photographed under shortwave ultraviolet light, effectively illustrates why the Franklin-Sterling Hill area is so popular among collectors of fluorescent minerals.

Plate 25. A vein of coarse-grained calcite (fluorescent red) and willemite (yellowish green) within fine-grained willemite ore from Franklin. The vein is about one inch thick and is shown here under shortwave ultraviolet light. Mitch Dolobowsky photo.

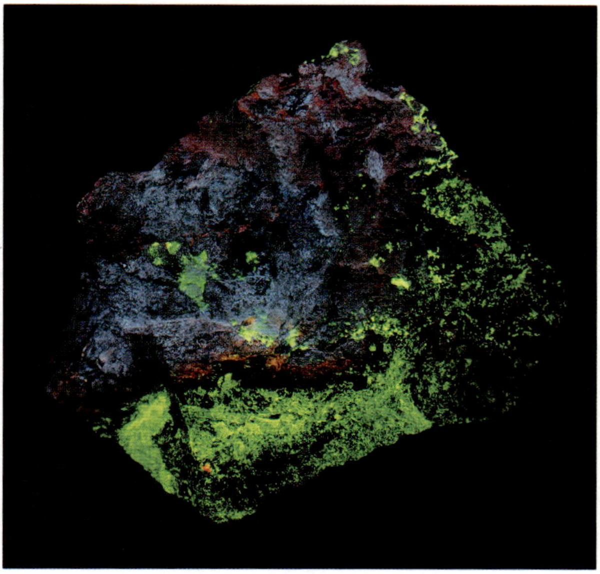

Plate 24

Plate 25

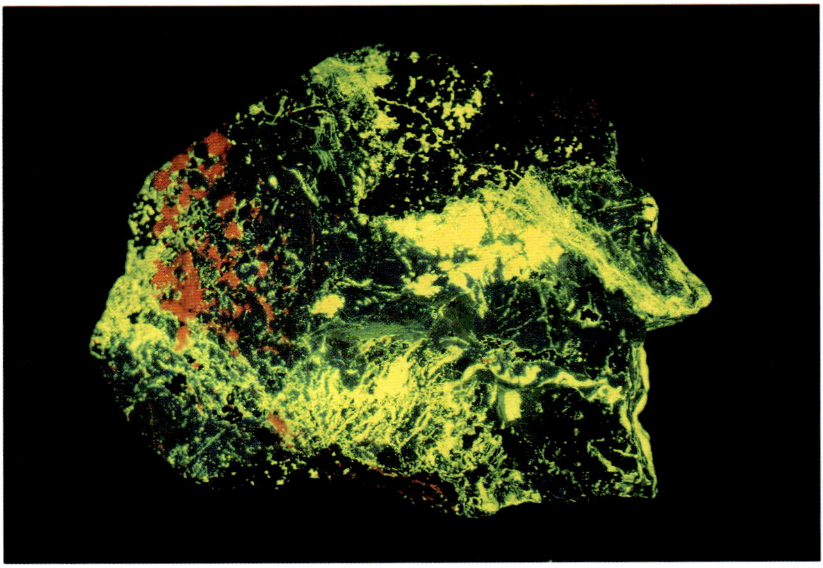

Plate 26

Plate 26. A four-color fluorescent specimen from Franklin, as seen under shortwave ultraviolet light. Shown here are esperite (fluorescent bright greenish yellow), willemite (yellowish green), calcite (red), and hardystonite (deep violet-blue).

Plate 27. Diopside in pale blue-fluorescing grains richly disseminated in nonfluorescent calcite, from Franklin. The specimen, shown here under shortwave ultraviolet light, is about three inches across. Henry Van Lenten photo.

Plate 28. An irregular vein of secondary zincite cutting willemite-calcite ore from Sterling Hill. Under the longwave lamp, as shown here, the fluorescence of the three minerals is in perfect balance. The zincite vein is about one-half inch thick. Henry Van Lenten photo.

Plate 27

Plate 28

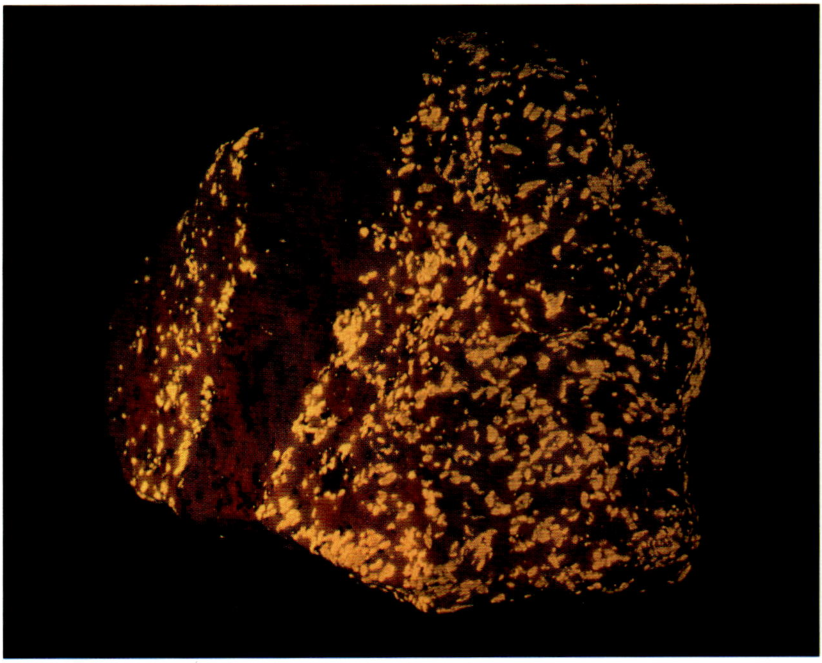

Plate 29

Plate 29. Grains of bright yellow-fluorescing wollastonite richly disseminated within dull red-fluorescing calcite from 740 crosscut on 700 level at Sterling Hill. The specimen is about five inches across and was photographed under shortwave ultraviolet light. Alfred Standfast photo.

Plate 30. A polished slab about seven inches long of willemite (fluorescent yellowish green) in calcite (red) from the 750 level, Franklin. The streaked appearance of the green-fluorescing grains is due to numerous, almost microscopic veinlets of secondary willemite cutting the original, more dimly fluorescent willemite grains. David Grigsby photo.

Plate 31. A selection of polished cabochons of the zinc ores from Franklin and Sterling Hill, as seen under shortwave ultraviolet light. These gemstones, the largest of which is about 1.5 inches long, illustrate the almost infinite variety in pattern and relative abundance of three of the most common minerals from this famous mining area: willemite (fluorescent yellowish green), calcite (orange-red to red), and franklinite (nonfluorescent). The tiny areas showing deep violet-blue fluorescence in several of the stones are hardystonite.

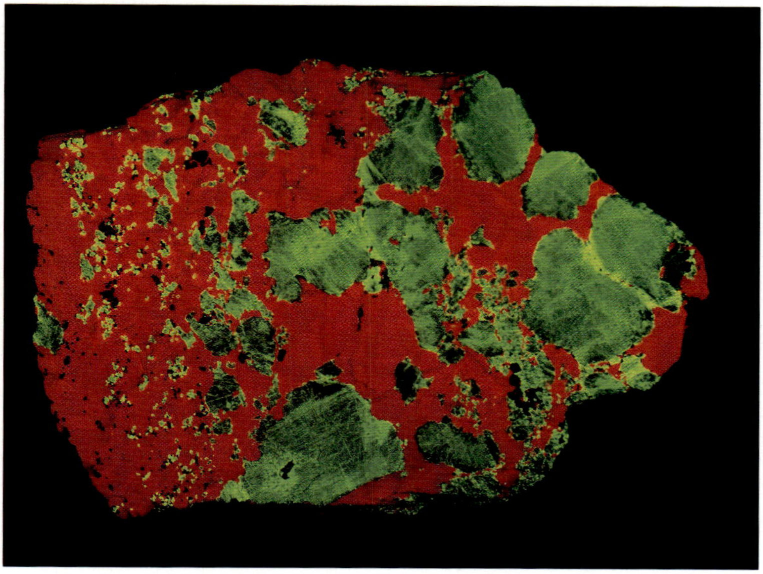

Plate 30

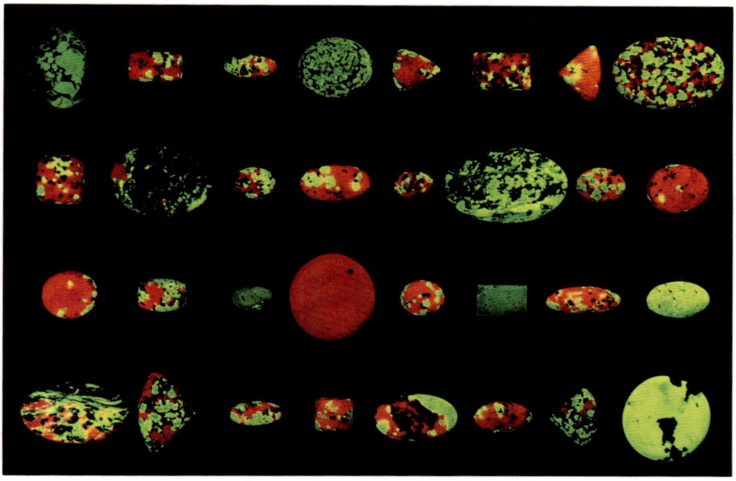

Plate 31

The Fluorescent Mineral Hobby

Richard C. Bostwick

Introduction: The Fascination of Collecting

Collecting has to be a universal characteristic of mankind. Everyone has the urge to possess what fascinates them, and each of us is fascinated by some aspect of the world. As children, most of us collected one thing after another as it struck our fancy: bottle caps, bugs, seashells, dolls, etc. The fact that most of these accumulations were thrown out as we grew up should never lead us to undervalue or ignore the impulses which led us to collect in the first place. Everyone who has ever brought home a pebble from the beach or a flower from the meadow, saying, "Look at this! I found it. Isn't it wonderful?" knows quite well what collecting is about.

Fluorescent minerals are, of course, remarkably beautiful. Their colors are vivid and striking under the ultraviolet lamp, and the contrast with their appearance in daylight is both dramatic and mysterious. They compel wonder. It would be difficult *not* to want to collect them.

Collecting fluorescent minerals, then, is a matter of acquiring beauty, enjoying it, and sharing it. Fluorescent minerals are exciting the first or the thousandth time you see them, and your excitement and enjoyment will be easily communicated to others. And there are advantages common to other hobbies, such as the fellowship of collectors with the same interests and enthusiasm. You can also indulge your hobby to the maximum degree allowed by your time, energy, and pocketbook. Soon you will find yourself building a collection which is unique in its scope and perspective. And with your acquisition of specimens you will also be gaining knowledge: a precious commodity worth even more when you give it away.

Collecting fluorescent minerals is an absorbing, multifaceted hobby. It involves finding minerals in the field as well as pursuing specimens from other sources. It means learning how to build and maintain your own unique collection. It certainly entails sharing the beauty of your minerals through displays at home and at mineral shows. Your collecting always will be a process of self-education and the education of others; the more you learn about fluorescent minerals, the more important it will be to pass along what you know about this extraordinary aspect of Nature.

Field Collecting

More than thirty years after Sterling Gleason wrote *Ultraviolet Guide to Minerals,* the attractions of collecting fluorescent minerals in the field are still very real. New discoveries are still being made, and classic localities reopened. Since 1960 hundreds of articles and notes have appeared about fluorescent minerals and where they can be found, so there is much more information available today for interested collectors. And the thrill of having a mine dump or quarry tailings pile come alight underfoot has not aged.

What *has* changed, particularly in the eastern United States, is the availability of collecting sites and the attitudes of responsible mineral collectors toward field collecting. On the cover of Gleason's original *Guide*, a collector is poised on the edge of an empty, beckoning landscape with an ultraviolet lamp in one hand and a rock hammer in the other. In the distance a mountain ridge looms, without question concealing hidden treasures; there is no road or trail, just the assurance of a collecting adventure. There isn't a "No Trespassing" sign in sight. There are no fences, no parking lots, no strip malls, no condominiums, and no insurance companies; in short, for most of us it's Fantasyland. Field collectors today are likely to find legal and physical barriers in their way, and field collecting has become almost a conservationist activity, with strong emphasis on not only personal safety, but also respect for other collectors, property owners, and the collecting locality itself.

Today there are really three essentials for responsible collecting: good protective clothing and equipment, safety-conscious behavior, and conservation of specimens and collecting localities. And because there is no substitute for experience, it is strongly recommended that beginners join local mineral clubs and have their first field trips in the company of experienced collectors.

Protective Clothing and Equipment

Safety goggles are the most important article of clothing in the field. Even when breaking rocks is not necessary, few collectors can resist the temptation. High-velocity rock chips often are the result; protect your eyes! Safety goggles are doubly important to collectors of fluorescent minerals as they also block shortwave ultraviolet energy; more on this later.

Sturdy footwear is also important. Mineral dumps are often unstable, and sneakers or sandals won't protect against jagged rocks or prevent twisted ankles. Good leather boots with lug soles are recommended, and steel-toed safety boots may be preferred for their extra margin of safety. Durable cloth or leather work gloves help to avoid damage to the hands, and long-sleeved shirts and long pants afford additional protection, not only from rocks but also poison ivy and insects.

On any trip to a mine or quarry, indeed to any place where you might encounter loose rock, overhangs, or low ceilings in passageways, a hard hat is necessary. Many mineral clubs make goggles, boots, and a hard hat required equipment for all field trips. Most field collectors keep these items in their automobile trunks along with their packs, hammers, chisels, pry bars, etc.

Speaking of hammers, chisels, and other collecting tools, quality is important. Steel tools from a reputable manufacturer are unlikely to break or chip, ruining a field trip or causing an injury.

Rocks are heavy. Most collectors do not realize how heavy until they have carried their prizes out of a deep quarry, down a steep mountainside, or up a mine ladder. Suffice it to say that a good backpack or even a strong canvas bag is bound to help. When the path to a collecting site is more or less level, many collectors also bring express wagons, hand trucks with buckets lashed to them, or other improvised wheeled conveyances. Here again, durable equipment is important; it is embarrassing and exhausting to have to carry out both your mineral plunder and your broken grocery cart or kiddie stroller.

With your goggles, boots, hammers, and packs, please do not forget wrapping materials and a notebook, to protect and label your find. Any specimen worth collecting is worth preserving free of dings, bonks, and whacks, which are exactly what a rock will acquire when thrown naked into a pack with other rocks for company. Make sure your wrapping material is free of fluorescent dyes; these can rub off on a specimen, endowing it with some unusual fluorescent properties. (If you must use newsprint, first cover the specimen with plastic wrap.) The notebook is for recording the specifics of the find: date, locality description, mineral associations, etc. Pages can also be torn out and used for makeshift labels to identify particular specimens. As fresh as the details of a mineral discovery may be at first, they fade with time unless written down. Most collections have a distinct proportion of mystery specimens because this simple step was omitted.

As a field collector of fluorescent minerals, you have two special needs: the first is a portable source of ultraviolet energy, and the second a means of creating enough darkness to view the fluorescence of your newly found minerals. As for the first need, portable battery-powered ultraviolet lamps are easily available. Some are small enough to fit in a pocket; though inexpensive, these lamps are surprisingly effective. Larger portable ultraviolet lamps have more powerful batteries and tubes, and even built-in flashlights. On a really dark night, these lamps can cause visible fluorescence in some minerals at a distance of twenty feet or more.

As for the second need, the portable equivalent of a dark corner is necessary because even the brightest mineral fluorescence is dim in daylight. Many collectors take a black plastic sheet with them; it is lightweight, easy to fold and carry, and effectively excludes much light when draped over one's head. Blankets, ponchos, etc., can also be used. The spectacle of a black blob wiggling over the surface of a mine dump may inspire some hooting, but the collector generally has the last laugh.

More sophisticated viewing devices for the field can be made or purchased. Among these is a commercially available "Viewing Bag." This has a collar and cuffs which exclude light; flipped over one's head, it isolates eyes, hands, specimens, and an ultraviolet lamp in a pocket of darkness. The bottom unzips so that the "Bag" can be dropped over a large specimen in the field or a box of specimens at a mineral show. There are also many variants of the portable viewing cabinet. In its basic form this is a container with two slots in it, one for an ultraviolet lamp and the other for looking at the minerals within; a cardboard box or plastic bucket with two rectangular cutouts will do in a pinch. More elaborate versions may have a contoured eyepiece, clamps to hold the ultraviolet lamp, curtains or trapdoors for moving specimens in and out, etc. Some cabinets are collapsible, some have removable bottoms; the features are limited only by one's ingenuity. Larger, rigid viewing cabinets can also be carried to the field in the back of a vehicle; specimens are taken to the cabinet for examination.

Field Trip Behavior and Conservation

What does "safety-conscious behavior" mean in this context? Primarily this: that on field trips you should be conscious of your own and others' safety. For yourself it is a matter of common sense: not walking over cliffs or into mine shafts, and remembering to wear your goggles when pounding away on a boulder. When others are nearby, it means thinking before you act. Don't roll rocks on them; don't swing hammers when they're near; and don't leave booby traps such as deep holes or undermined banks for the next collector.

Why would anyone, you ask, be so inconsiderate? Because they are. Many mineral collectors suffer from a disorder which might be termed "Collector's Myopia." It is very similar to "Buck Fever" among deer hunters and consists of being so caught up in the excitement of collecting that you become oblivious to everything and everyone else. As a safety officer on many field trips into working quarries, I have repeatedly been amazed how collectors will stray onto crumbling cliff edges or under dangerous overhangs minutes after they have been warned against doing exactly those things. They're not consciously posing a risk to themselves or others, they're following their bliss. Please remember when collecting that health is more precious than the finest mineral specimen.

On the rare occasion when a collector breaks a leg or leaves a sledgehammer in the rock pile where it will damage the quarry's crusher, he is generally ignoring the fact that all collectors will suffer as a result. Many prime collecting sites are now off-limits for everyone because of the carelessness of one individual. When there is an accident, even if the property owner is tolerant his insurance company will almost certainly drop him or increase his premiums drastically. This is where good manners and common sense work together to keep localities open for field trips by not alienating property owners. For example, permission to collect should always be obtained; you should never damage fences, machinery, buildings, etc.; and you *should* fill in holes and take your own trash out with you.

How is the mineral collector a conservationist? By avoiding abuse of the land, by not overcollecting and exhausting unique resources, and by behaving in such a way that future collectors will be welcome. Greedy and destructive treatment of a mineral locality not only discourages other collectors and aggravates property owners but also destroys scientific evidence. Mineral occurrences are best studied in place; it is very difficult to reconstruct the overall geology of a locality from hand specimens. And again, permission to collect is very important. In the United States, for example, all caves are off-limits and road cuts (however tempting) are always questionable. Mine sites and quarries are more likely to be open to qualified individuals or groups. Keep in mind that occasional supervised access to a mineral locality is infinitely preferable to a brief orgy of irresponsible collecting which closes that locality for good.

Collecting After Dark

Alone among mineral hobby enthusiasts, collectors of fluorescent minerals do their best work at night. Nocturnal field trips are always more interesting and productive than ones held in broad daylight. However, be aware that it is often difficult to talk either mineral clubs or property owners into sanctioning such activities. When and where it is possible, night collecting is perhaps the supreme thrill for the initiate. There is much to be said for an outdoor mineral display glowing as far as your ultraviolet lamp will reach, and in the best cases this is what you may find.

The same cautions apply as in collecting by daylight: wear protective clothing, get permission to collect, and don't be greedy or careless. Remember, too, that local flora and fauna can be hazardous to your health: for example, in the East, poison ivy; in the West, scorpions. The one nice thing about scorpions is that they fluoresce vivid greenish-blue. Don't be too impulsive in picking things up because they glow.

Any potential after-dark collecting locality should first be scouted in daylight, not only for fluorescent minerals, but also for dangers which might not be obvious in darkness. Mark any hazards clearly with fluorescent plastic ribbon, which is available from industrial safety suppliers, or with fluorescent chalk. (The ribbon may be more difficult to obtain, but ribbon can be taken with you when you leave and used over.) Also mark all tools, lamps, bags, etc., with fluorescent tape or paint or felt-tip marker; otherwise they can be very hard to find once you have set them down. A flashlight should *always* be taken, along with replacement batteries and bulb; a spare tube and batteries should also be taken for each ultraviolet lamp. A further refinement is a stiff metal screen taped over your ultraviolet lamp's filter. This prevents an abrupt and expensive end to a field trip in the event that you drop the lamp filter-side down or bash it into a rock you are inspecting. Finally, always allow your eyes to adapt to darkness before getting underway. The vivid colors of fluorescent minerals in their natural setting should not be dulled through eagerness; you are also less likely to walk off a cliff.

Finding More Mineral Localities

Advanced collectors of fluorescent minerals may find that they have exhausted the resources of their local mineral collectors and organizations. Here are some approaches which can help you to find new fluorescent mineral localities or rediscover old ones. One excellent source of information is general mineral collections. Prior to roughly 1940 almost no one collected fluorescent minerals; since then not many mineral collectors have concentrated on the subject. Consequently, a few hours spent in the dark with an ultraviolet lamp and an older collection can bring many things to light, such as unfamiliar or exceptional fluorescent minerals from both classic and forgotten localities.

Publications can also be rich sources of information: mineralogical textbooks and reference works; periodicals for mineral collectors (e.g., *The Mineralogical Record, Rocks and Minerals, The Mineralogist*) and those for mineralogists (e.g., *The American Mineralogist*); and geological survey reports and topographic maps. For example, a turn-of-the-century U.S. Geological Survey report about a mining area will include the names and locations of mines, prospect pits, and quarries now overgrown, obliterated, or simply forgotten; current updated topographic maps often lack such information.

Furthermore, many localities for fluorescent minerals remain undiscovered or unreported, because most prospectors and geologists never bothered to check for fluorescence. Be prepared. Always have a portable ultraviolet lamp with you in your travels, and make a point of exploring mine dumps, road cuts, quarries, cliffs, and outcrops. The potential for discovery is high, much higher than for minerals which are obvious in daylight. Fluorescent minerals are very much where you find them.

Becoming A Mineral Collector

You don't become a mineral collector just by picking up a pretty mineral, although the impulse to acquire beautiful things is at the root of collecting. Real collectors are always curious about the objects of their desire and never stop learning about them any more than they can stop picking them up. As your own collection grows, you will discover that your first-hand knowledge of minerals must be balanced by book-learning, and that both are necessary for full appreciation, understanding, and enjoyment of your hobby.

Acquiring Knowledge

Collecting without knowledge is mere accumulation. There is much information available about fluorescent minerals: what they are, why they fluoresce, where they can be found. Some of that information is in this book, but much more is available through collectors, libraries, and mineral organizations. Eventually you will be able to contribute to that specialized body of knowledge; in the beginning you need to know where to find it.

The first thing to do is to join your local mineral club. This will put you in touch with collectors from whom you can learn an enormous amount: not just information about minerals in general, but also collecting localities and field trips, the mineral dealers in your area, the times and places of mineral shows (ranging from swap sessions to large, important exhibitions), the nearby museums with good mineral exhibits, and so on.

The next thing is to learn as much as you can about your hobby. There are numerous mineral magazines, some of them available at newsstands or bookstores. Read them, subscribe to them. See what your school or public library has under Earth Science, Geology, Mineralogy. If your mineral club is a member of the AFMS (American Federation of Mineralogical Societies), ask what resources are available to your club through the Federation. These may include such things as slide shows about fluorescent minerals, and the full range of AFMS topics may include many others which interest you. As you progress, you may find that a local college or university has an Earth Sciences Department with an extensive library and quite possibly students or professors with an interest in fluorescent minerals.

Sooner or later (preferably sooner) you will join the Fluorescent Mineral Society. This is a specialized mineral club with members throughout the world, linked by a newsletter (*U.V. Waves*) and the *Journal of the Fluorescent Mineral Society*. From the *U.V. Waves* you will obtain lists of members, offers to trade, notices of F.M.S. activities, and much other information. The *Journal* has many articles of lasting value. The 1988-89 issue, also known as *The Henkel Glossary of Fluorescent Minerals,* is a list of almost all minerals which have been reported to fluoresce: 566 species (!) in all, plus assorted rocks, organic substances, etc. It is a valuable reference work for every collector of fluorescent minerals. More recently an Advertising Supplement has been published which combines articles of general interest to collectors with ads for mineral dealers and for suppliers of ultraviolet lamps and accessories.

Depending on your walk of life and particular interests, you may find that you can make unique contributions to the body of knowledge about fluorescent minerals. There is much serious basic research waiting to be done about them; their physics and chemistry are still imperfectly understood, for example. Furthermore, there are many projects in which the Fluorescent Mineral Society's members are involved that are likely to be ongoing for many years, such as understanding the solarization of shortwave filters, enlarging and improving the *Henkel Glossary,* and producing a Fluorescent Color Standard.

The area where a collector of fluorescent minerals is most likely to be able to give something back to his hobby is in the description of individual mineral localities. There are hundreds of these for which nothing has been written down, small localities which are known well only to collectors but which have produced some remarkable fluorescent minerals. There are also many better-known localities which have been described extensively but with scant attention to the fluorescent minerals found there. The potential for contributions of this sort by mineral collectors is very great.

Acquiring Specimens

Important as information will be for you, you cannot have a mineral collection without mineral specimens. Initially, you will find that fellow collectors are often happy to share their minerals as well as their time and knowledge. As you develop your own taste in minerals, however, you will certainly acquire most of your specimens through field collecting, trading, and purchasing.

Field collecting has been covered earlier, but the fact remains that regardless of the volume or quality of the specimens you find, the experience itself is crucial. Seeing minerals in their natural habitat is as important for understanding them as it is to see animals in the wild. Learning the techniques and difficulties of field collecting will also help you to appreciate how rare and remarkable good specimens are, and why we should work hard to conserve our mineral heritage.

Trading often begins with specimens you have collected yourself. Almost every collector will find that there is at least one type of material which is abundant locally and scarce elsewhere, and good collections have been built by simply exchanging one region's surplus for another's. Trading can be local, national, or international, face-to-face or through correspondence. Exchanging specimens with collectors in other countries is often the best way of obtaining rare and desirable material, as currency control laws and bank regulations abroad can make it both difficult and expensive to send or receive money for specimens. Moreover, money to some collectors and museums has much less importance than a mineral their collection lacks. This is as true in the United States as elsewhere.

However, for most collectors there is really no escaping the use of cash, the "silver pick". A generation or two ago there were many collectors who rarely bought or sold specimens, taking pride in their field collecting and trading skills. Today the exhaustion or closure to collecting of many mineral localities, coupled with the increasing amount of material going into collections, has led to greater commercialization of the hobby. Many of the fluorescent minerals now in circulation are from collections, and most collections are sooner or later sold to dealers, then dispersed. One interesting result of this is the growing availability of fluorescent mineral "classics," beautiful specimens from localities exhausted long ago. Initially these disappeared into collections, but now that dealers are recirculating them the average collector has a much greater chance of finding that rare, key "old-timer." It is also true that the "silver pick" gives us access to localities which never were open to collecting, and to others worldwide which would be incredibly expensive to visit.

For the beginning collector in particular, a true attraction of fluorescent minerals is that they are relatively inexpensive. Well-crystallized specimens that are richly colored in daylight often cost a good deal. After you have seen what is displayed in museums, and what is charged for similar pieces at mineral shows, you will begin to understand what can be involved in assembling a major collection of fine mineral crystals. Collectors of fluorescent minerals simply do not face obstacles of that magnitude. In the Franklin-Sterling Hill area, for example, there are dozens of private collections with displays of the local minerals which are the equal of anything to be seen in museums in New York City or Los Angeles. The same is true for private collections of worldwide fluorescent minerals throughout the United States and abroad. Simply put, every collector of fluorescent minerals has the opportunity to build a good collection of attractive specimens without first becoming rich or going into debt. The greatest expense a beginning collector will have is that most necessary of tools, a good ultraviolet lamp. And here it must be said that the better the lamp, the more enjoyment your collecting will bring you and others. The finest fluorescent specimens in the world will excite no one if their fluorescence cannot be clearly seen.

If you love fluorescent minerals, collecting them can be an important part of your life. Some simple advice may be in order. First, get to know the collectors and dealers in your area. In mineral collecting, as in other hobbies, the fine line between the dealer and the collector is often a tax-exempt number. Sooner or later, as your collecting tastes and concerns evolve, you will find yourself with a surplus which may be traded (as outlined earlier) or sold. In any community of mineral collectors there is constant circulation of material. The better you know this community, the easier it is to participate. "Wheeling and dealing" may not be to everyone's taste, but it is an excellent way to upgrade your collection and continually refine it. Often mineral collecting is as much a social activity as an economic one, and it is best to recognize this.

Now seek out dealers and collectors through the mail; it is a larger community where many of the rules learned locally will apply. You will soon learn who shares your interests and tastes, and these relationships are rarely one-sided. Collectors of fluorescent minerals are so widely scattered that they and the dealers they patronize do much of their trading and buying by mail.

Be aware that many mineral dealers who do not make a specialty of fluorescent minerals will still have some in stock; they may be willing to let you go through that stock with your ultraviolet lamp. This is true even at mineral shows. There will be a dark area under a table or around a corner, and if you have brought your portable ultraviolet lamp, you can get some idea of how well a specimen fluoresces. (Some collectors and dealers will even bring to shows portable viewing boxes or bags such as might be used during field collecting.) Please remember to ask a dealer before handling a specimen or removing it to a dark corner. The less well the dealer knows you and the more fragile and valuable the specimen, the less likely the dealer is to cooperate with your enthusiasm.

When acquiring minerals, your best defense is your knowledge—yet another reason for learning as much as possible about your hobby. The Latin adage Caveat Emptor ("Let the buyer beware") applies as well to mineral collecting as to other endeavors. In their defense, nearly all dealers are honest; their reputations are important to them, and they are sensitive to criticism. But the more you know about fluorescent minerals, the less likely you are to be disappointed in any kind of a "deal."

Finally, please remember that ANY hobby should be pursued for pleasure, not profit or pride. It is a mistake to let your hobby-horse ride you. If it happens that your collection with time becomes valuable, or that part-time dealing enables you to upgrade specimens while staying within your budget, well and good; this happens to many. However, if obsession with money or status supersedes your fascination with minerals, you will have lost the enjoyment of beauty which drew you to your hobby in the first place.

Maintaining A Mineral Collection

As you collect, information always arrives along with each specimen: when and where it was collected, and by whom; its price, if purchased; and possibly an identification, either yours or someone else's, of the mineral species present. If an older specimen, it probably has a history of collections it has passed through, either recorded on older labels or passed along verbally. Of all this information, only the identity of the mineral(s) is inherent in the piece, because it can always be retrieved by various mineral identification techniques. ALL the other information stands to be lost if the specimen is treated carelessly.

You are the curator of your own collection. As such, you are responsible for conserving both specimens and information. Preserving that information in such a way that it stays with the specimen is critically important. Many collectors, because they can remember all the details about a specimen, think there is no point in writing down any of them. Their collections are neither labeled, numbered, nor logged. Such a collection, when passed on without documentation, loses much of its true value forever. If that information is not written down, it *will* be lost.

Why is this information important? Because the Earth has a history, and the efforts of geologists and mineralogists are directed toward understanding it. The history of an individual specimen is directly linked to the history of the Earth through a complex web of collectors and scientists, and their activities. Let us say that you are a scientist writing a comprehensive paper about a long-exhausted orebody; by studying its mineralogy, you hope to understand how it originated, how it developed, and how its history is related to the geology of the region. The site of the mine and its dumps is now covered by a housing development, and all you have to study is specimens from that mine. How wonderful if there were hundreds of specimens, each labeled with the date it was found, the level and working place in the mine, and the person who found or saved it! From this you could reconstruct much besides a list of species for the deposit, including not only the relative position of different mineral suites in the orebody, which could tell you a great deal about origins and development, but also something about the history of the mine and its miners, which might in turn contribute to an understanding of the area's cultural and economic history. If instead you found only a handful of specimens which might or might not have come from that particular mine, how much responsible work could you do?

Please learn to conserve information along with the specimens themselves. Here are some suggestions:

Your Mineral Catalog

Keep a log, preferably in a permanently bound notebook. Record from whom you acquired each specimen, and when, and the price paid. If the history of the specimen is known from old labels or by word of mouth, include that. Write down the mineral species known or believed to be present. Above all, record the locality as exactly and completely as possible. And give each entry a number which will also be marked on the specimen itself.

In numbering you may wish to prefix the number with a letter or letters, usually your initials. This will mark the specimen as a cataloged item in your collection. In the event the specimen is misplaced, sold, or traded, or the collection is broken up, at least specimen GW125 (for example) is still traceable to the George Washington collection and to its logbook, if that is preserved.

The name of each mineral species should be spelled correctly, and distinguished from commercial or varietal names. Every mineral collector should own a few basic works to help him or her understand minerals, and the basic reference for species names is the *Glossary of Mineral Species* by Michael Fleischer and (as of the 1991 Sixth Edition) Joseph Mandarino. It is generally available through mineral dealers and mineral magazines.

When recording a locality in your log, be as specific as possible. If the specimen is from a mine, include the level and working place, mine coordinates, or whatever information is available. For open-pit mines and quarries, there may be similar details (e.g., north end of quarry, second bench). If you know the approximate date the specimen was collected or who collected it, write that down. If the locality is known by more than one name, include them as well. As fresh and familiar as such information seems to you now, it will evaporate if you don't record it, and future collectors and scientists may value it more than you can imagine.

Correct locality information, while extremely important, can be hard to pin down when you are cataloging an older specimen. Old locality names can be particularly treacherous: as time passes, the names of towns and districts can change, while a local mine, quarry, or collecting site can not only change names but can also be referred to in more than one way during a given period. European mineral localities are notoriously subject to name-changing, as many have histories of hundreds of years during which borders have shifted, countries have been broken apart and others created, and local languages and ethnic domination have varied. When faced with an informational morass like this, never be afraid to reach out for help, as there are excellent articles in periodicals, sections in mineral reference works, and other sources of assistance, notably a good atlas and/or gazetteer.

If you are a field collector, precisely locating a prospect pit, a section of a quarry, an old mine dump, etc., can be difficult, but it is worth the effort to get it right. Perhaps the easiest way is to mark the spot on a topographic map, if one is available. Detailed maps for much of the United States are available from the U.S. Geological Survey, and many suppliers of outdoor recreational equipment carry the same maps. If a map is not available, avoid describing a locality in terms of temporary landmarks such as fence posts, trees, or barns. With a compass you can locate a point by taking the compass bearings of two or more permanent landmarks from that point. Beyond this there are other accurate locating methods; suggestions can be found in an article by Pete J. Dunn in *Mineralogical Record,* vol. 22, no. 5, p. 330 (1991).

For a large and growing collection, consider computerizing your catalog; there are many advantages to this. See the article by Tony Kampf in *Rocks and Minerals,* vol. 64, no. 2 (1989).

Labeling Your Specimens

Label everything. Even bulk specimens from recent field trips should at least be segregated in a box marked with the locality and the date collected. The shallow cardboard trays in which cases of canned soft drinks and beer are sold are recommended for this; some sizes act as lids for others, and such "rock boxes" are easy to find, pack, lift, and store.

Once sorted, every specimen placed in your collection should be labeled or marked in some way. The conventional approach is to keep each specimen in its own box or tray, with a paper label underneath the specimen. If a specimen is wrapped for storage or shipment, its label is wrapped with it. This has disadvantages, chiefly that the label can be easily damaged or lost, thereby severing the specimen from its identity. A more permanent connection should be made by marking the specimen itself.

If a specimen has been logged, the simplest way of identifying it is to write its catalog number with India ink or a fine-point permanent marker either on a dab of paint applied to the specimen or on a piece of paper which is then glued to the specimen. It is still possible for the logbook and the specimen to become separated, so a more satisfactory identification can be made by adding to the catalog number the exact locality for the specimen and (if space permits) its key mineral species. Of these the locality information is the more important; in fact, even if you do not have a logbook or catalog numbers, you should *always* attach locality information to a specimen. Species identity can be determined later from the specimen itself. When gluing a label to a specimen, keep the label small and place it where it will not disfigure the specimen, preferably opposite the best display face. With practice, neat handwritten labels can be made with India ink and a crowquill pen. Larger handwritten or typed labels can also be photoreduced to the appropriate size, and if you have access to a word processor with a laser printer, excellent small labels can be made with that.

As most specimens need cleaning sooner or later, labels attached to the specimen should be designed so that they do not wash off. Paper labels glued to the specimen should be varnished to prevent softening and disintegration when wet. Water-based glues such as *Elmer's* usually will not fluoresce but will hold through many washings.

When choosing a labeling system for your fluorescent minerals, pay attention to the way the labels will look under both daylight and the ultraviolet lamp. The specimen trays and labels used by most dealers are fluorescent, as is the cotton padding often found in these trays. If you make your own labels or have them made up by a commercial printer, choose paper which fluoresces poorly or not at all. For labels painted on or glued to the specimens, avoid glues, papers, or varnishes which fluoresce brightly; also avoid bleached white paper, which may deteriorate with time. Acid-free archival grade paper is preferable, and this may be obtained from good stationery stores or perhaps through libraries. An inexpensive and effective substitute is the brown Kraft paper used in paper bags. It is unbleached and will not deteriorate.

Taking Care of Your Minerals

Treat specimens carefully, so that they don't get damaged or separated from their labels. One of the most discouraging sights a mineral collector can see is a box of loose specimens bashed beyond recognition, with their labels in tatters.

A collection can be kept in shallow cardboard boxes (as mentioned earlier) with the specimens in individual specimen trays, but it is more satisfying to have them in a less undignified setting. Cabinets made for minerals have shallow, strong drawers, preferably of wood. Reasonable substitutes may be found in antique stores (e.g., spool cabinets, which are popular for holding miniatures and micromounts) and stores which sell secondhand office furniture (e.g., map or blueprint cases and file card cabinets).

If displayed in the house, minerals should be protected from dust, cooking or tobacco smoke, and direct sunlight. Some minerals are also sensitive to heat and humidity. If stored in a basement or outside storage shed, minerals should be wrapped and boxed; fluctuating temperatures can cause condensation of water on exposed mineral surfaces, with serious consequences. For example, most Franklin and Sterling Hill specimens will eventually turn brown under these conditions.

When wrapping or padding mineral specimens for temporary transport or long-term storage, watch out for materials which are fluorescent and may leave a residue on the specimen. The chief offenders are the cotton pads from specimen trays, and paper dyed with fluorescent dyes. Ink and fluorescent dyes in newsprint can rub off on minerals. Further, newsprint and other papers can deteriorate with time and release chemicals which can damage specimens. Brown paper or plastic are the best wrapping materials for minerals.

Cleaning Your Minerals

Related to protection of mineral specimens is the matter of cleaning them, which is NOT simply a case of plunging them into hot water with some dish detergent and scrubbing them vigorously. Some cleaning agents fluoresce, while some minerals are damaged by water and others are too fragile to scrub.

Most detergents contain optical brighteners which fluoresce bright blue-white under longwave and shortwave ultraviolet light. If a specimen has been washed in such a detergent, the brighteners can be difficult to remove. It is better to wash minerals with a pure soap such as Ivory, which is easier to rinse completely off a specimen and does not contain fluorescent dyes.

Some minerals are actually water-soluble. An example is halite, which is rarely found as crystals and crusts that fluoresce bright red under shortwave ultraviolet light. Essentially halite is table salt; if you wash it in water, it dissolves. Other minerals react more slowly to water and dampness but may crumble or develop unsightly crusts; such minerals are often stored with a desiccating agent or sprayed with a waterproof coating to preserve their appearance. While such practices are common for certain minerals, collectors of fluorescent minerals should be aware that these coatings can fluoresce, or mask the fluorescence of the minerals beneath. Further information about minerals which are affected by water may be found in Richard M. Pearl's *Cleaning and Preserving Minerals* (Earth Science Publishing Co., Colorado Springs, CO; various editions).

Many minerals are easily damaged by scrubbing because they are soft and/or form delicate crystals. Hydrozincite, a collector's favorite with its bright blue fluorescence, occurs as soft crusts and as druses of small, fragile crystals, either form vulnerable to a toothbrush. Mineral specimens do get dirty and have to be cleaned, but you should always be sure that the patient is not killed by the cure.

Displaying Fluorescent Minerals

In many ways this is the crux of the hobby. Talking about the beauty of fluorescent minerals will convince very few people; you have to show them. And a good display seen under proper conditions is breathtaking: bright, sharp colors standing out in the dark, like silent, frozen fireworks.

People seeing mineral fluorescence for the first time will ask "What makes those colors? Why are they so bright?" and other questions, including the inevitable, "Are these rocks painted?" Explanations can come later; the point is that the appeal of fluorescent minerals is immediate and direct. To maximize this appeal, here are a few simple rules for displaying:

Basic Principles of Displaying

1. *The display area should be as dark as is practical.* The darker it is, the brighter the fluorescent minerals will appear. However, a balance should be struck between the brightness of the minerals and the needs of the audience. Most people coming to your display, whether it is at home or at a mineral show, will be walking in from a well-lit area. Their eyes will not be adapted to the dark, so that a plunge into absolute darkness (except for the light from the fluorescent minerals) may be unusually disorienting. Spectators need to see enough of the room to be comfortable there; if you can, give them a minute or two for their eyes to begin to adapt.

Although light sufficient for navigation should be present in the display room, a minimum of light should fall on the display itself. Fluorescent minerals are not so bright that they can compete with daylight or strong artificial light. Expecting someone to enjoy an exhibit of fluorescent minerals with light falling across it shows a lack of respect for the spectator and the minerals as well.

If there are windows near the display and the light coming past the window shades or curtains reaches the display, you can always block that light with a black plastic sheet held in place with duct tape. This is a standard technique used at mineral shows to prepare a display area, and one that is quick and simple. However, before using tape, you should make sure that it will not disfigure paint or wallpaper.

Keep in mind also that the area around your display should be free of clutter. Even with low-level lighting, your audience may be distracted by the beauty of the minerals, and it will detract from their experience to have them colliding with chairs or tripping over coffee tables.

2. *The backdrop for your specimens should be black or at least nonfluorescent.* You simply don't want any extraneous light to distract the viewer. Whether the minerals are arranged on bookshelves or a tabletop, or kept in a china cabinet or a made-to-order case, you will want to avoid fluorescent woods, paints, varnishes, cloths, etc. You will also find it advisable to vacuum the display from time to time, to pick up fluorescent lint and mineral debris.

It is best if the background for a display is coated or lined with some nonreflective black material, whether paint, cloth, or paper. Glossy surfaces will reflect both mineral fluorescence and the purple light from the ultraviolet lamp. For this reason alone, glass shelves (often preferred in mineral cabinets for daylight displays) are not recommended. Glass fronts for cabinets, which come between the viewer and the minerals, are another matter; they have their pros and cons, which will be discussed later.

3. *The ultraviolet lamp should always be held or positioned for maximum fluorescence.* If you are using a handlamp, the display can be of any size, but you should hold the lamp close to a few specimens at a time and then move on to the others. If you have a fixed display lamp or lamps, your exhibit should be proportioned to the area covered by the lamp(s). Matching the scale and type of your display to your ultraviolet lamp(s) will be covered in more detail under *Building Your Display* (p. 96).

Arranging Minerals in Your Display

It is said that there are no disputes concerning individual taste, which means that no one can tell you what you do or don't like. However, there are some general rules about placing specimens in a display to improve the chances of everyone liking it.

In the displaying of fluorescent minerals, there seem to be two types of display: ones which separate specimens and ones which cluster or push them together. The former, where specimens generally are arranged in rows with each piece distanced from its neighbors, is more common. Each piece stands by itself, framed in black and possibly accompanied by a label. Most competitive exhibits take this form. In the latter type of display, specimens are clustered or piled together, which gives a pleasant overall effect but makes it rather difficult to distinguish individual pieces. Both types of display have their own merits.

Displays which present specimens separately allow a viewer to concentrate on them one at a time. For any kind of a systematic exhibit, where a range of different minerals is being shown, this type of display is preferable. When setting it up, try for maximum color contrast between adjacent specimens. Avoid lumping colors that are the same or close by in the spectrum. For example, put red next to green or blue rather than next to orange or another red (*fig. 1*). Also avoid marked contrasts of specimen size or fluorescent brightness where possible. A rare, special piece which is neither large nor bright will be ignored if placed next to a brighter, larger example of a common mineral.

If your specimens vary in size and fluorescent color, but are of similar brightness, order them by size in your display. They can be ranked by row, with bigger specimens toward the back of the display, or ordered within a row with the larger pieces toward each end. If fluorescent brightness varies, the less-bright pieces should go toward the center of a row and/or the front of the display, to be closer to the viewer and the ultraviolet lamp. If you have a particular favorite which you want to be seen, put it front row center and consider elevating it on a small box or other mineral stand (*fig. 1*).

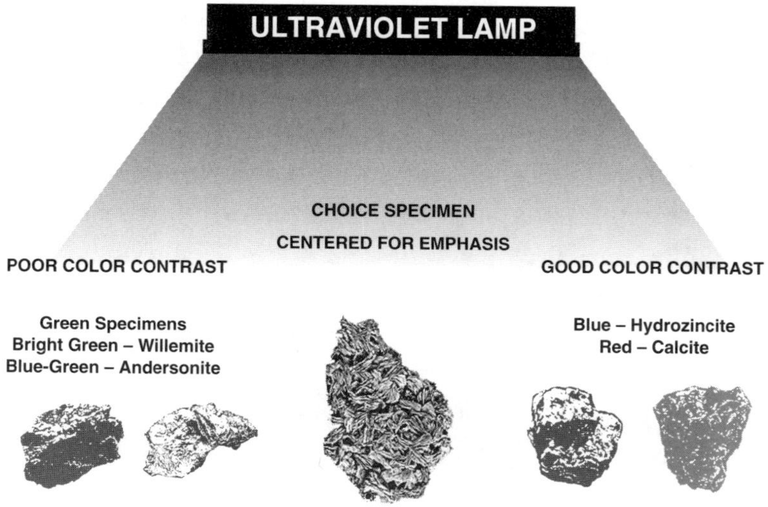

ARRANGING YOUR DISPLAY
FIGURE 1

Every specimen looks its best when seen from a particular angle. Part of the joy of putting together an exhibit is adjusting the position of each piece until it looks "just right." However, there will always be a few specimens that need to be propped up in some way. Any prop is legitimate if it neither distracts the viewer nor harms the specimen. Positioning aids typically include blocks of wood, pieces of cardboard, wire and plastic mineral stands, modeling clay, putty, etc. (Clays and putties should be chosen carefully, as they can stain a specimen or be difficult to remove; some also fluoresce.)

Positioning is less important in the alternate form of display where the specimens are massed. Here minerals are crowded together, touching, overlapping, and even piled atop one another. The most common type has the minerals arranged on a series of shallow steps, so that they appear as a roughly surfaced cascade, almost a rock waterfall. Many free-form variations exist; others wait to be created. Attempts have been made to imitate a rich mine dump, just as a collector might stumble across it, or to duplicate the monument of scheelite or other ore minerals which a prospector might build to mark a claim.

Although in daylight these massed fluorescent displays seem confused and cluttered, under the ultraviolet lamp they have a certain distinctive charm like that of an Oriental rug, balancing the well-defined small patterns of individual specimens against a looser overall design. It is a pleasure to let your eyes wander over such a bright hodgepodge, picking out unusual colors and striking effects. Massed displays also have the advantage of not being dependent on a few choice specimens; every piece enriches the whole.

Labeling Your Display

Public displays, as in museums and at mineral shows, generally require labels. Labels are optional in private exhibits whose owners are usually present to describe their collections. Whatever your taste in the matter, here are some suggestions. Remember that labels are intended to inform viewers, not distract them.

The simplest labels are those handwritten with fluorescent inks. Most stationery stores, and many department and drug stores selling school supplies, have felt-tip markers and pens identified as fluorescent, "neon," DayGlo™, etc. It is advisable to check these out with a portable ultraviolet lamp before buying them, as some do not fluoresce, and many have a daylight color different from their fluorescent color. Still, these markers are inexpensive, easily available, and effective. Art stores usually have a greater selection and are more likely to have fluorescent paints and inks as well.

Care must be exercised in choosing the heavy paper or thin cardboard stock from which display labels are usually made. Here again a portable ultraviolet lamp can be helpful. Avoid brightly fluorescent stock; often it is brighter than the minerals. Weakly fluorescent or nonfluorescent stock is preferable. Although labels have been made with nonfluorescent letters on a highly fluorescent background, they are not recommended. Such labels are easy to make because they can be typed, but in a mineral exhibit they are severely distracting.

The content of the labels will vary somewhat from collector to collector and from display to display, but all labels should clearly convey essential information. For an individual specimen, a label should include the species name(s) of the dominant fluorescent mineral or minerals, and the locality where the specimen originated. Usually such labels are small, about the same size as business cards. Larger labels (signs, really) are often necessary to provide the title for a display.

In competitive displays the rules for labeling individual specimens are explicit and strict, with penalties for errors of mineral nomenclature, locality information, and spelling. While no points are taken off for mistakes or lack of showmanship in a private or noncompetitive public display, it would be wise for all collectors to honor equally high standards of accuracy and neatness in *any* display. The world's best specimen of esperite will get little respect if it appears on exhibit with the label:

ESPURITE
FLOURESENT YELLOW
FRANKLIN, NY

Fluorescent mineral exhibits often present special problems in labeling, as when there are two or more fluorescent minerals in one specimen. Do you list all of them or just the dominant ones? If your audience (or part of it) won't know which mineral fluoresces which color, do you give this information as well? What if there is no dominant mineral? What do you do when you want to emphasize the number of fluorescent species? (To most collectors of fluorescent minerals, the more fluorescent minerals in a specimen, the higher its status.) Where do you draw the line between being thorough, accurate, and complete, and risking losing your audience through information overload? Many solutions are possible. The key point, however, is that giving out inaccurate, incomplete information is the greatest risk of all. If the specimen is worth looking at, it will take care of itself.

There are several ways of labeling specimens with multiple fluorescent minerals. One is to follow the name of a species with its fluorescent color:

WILLEMITE (green)
CALCITE (red)
FRANKLIN, N.J.

A second is to have the name of a mineral appear on the label in the same fluorescent color as the mineral itself. This can be technically difficult, but the results are pleasing. Yet a third solution is to make up a separate list, a key to fluorescent colors, posted inside the case. This works well only when a display (e.g., a case of Franklin, N.J. material) has a lot of duplication and few chances for confusion.

As mentioned, the simplest approach to making labels is to hand-letter each label with a fluorescent felt-tip pen. Drawn with care and practice, handmade labels can be attractive and effective. Lines can be ruled and stencils or templates used if one's unaided efforts need improvement. Labels are most often lettered in block capitals, but legibility is more important than the shapes of the letters. Italic script is also impressive if well done, and many felt-tip markers come with chisel points intended for this use.

There is still no avoiding the fact that labels which appear mechanically printed have a more uniform, "professional" look than those made by hand. Here are a few methods which require extra effort but yield superior results:

1. Use a word processor or computer with a laser printer to create labels with the appearance of a photographic negative, i.e. with everything black except the letters. Print these labels on moderately fluorescent paper. If necessary for stability, affix the paper to stiffer stock.
2. Make up conventional black-on-white labels and photograph them on black-and-white film. Develop the film and overlay the negatives on moderately or brightly fluorescent stock (paper, cardboard, plastic, etc.).
3. Have letters inscribed on laminated plastic sheets, the type commonly used for military and police name tags. Put fluorescent paint into the inscribed grooves forming the letters.

There are undoubtedly other ways of labeling fluorescent mineral displays. It is well to remember, however, that both the quality and the accuracy of your labels are critical. Nature has created minerals, but you alone are responsible for your labels. You will find that the time and effort put into accurate and attractive labeling are always worthwhile.

Building Your Display

The underpinnings of a fluorescent-mineral display are important. Minerals are too heavy to be hung on the wall and too bulky to be glued in an album. They should be well supported and shown to best advantage. Good displays can be built using existing structures, such as bookshelves, china cabinets, and closets, or by modifying ordinary mineral-display cabinets, or by building from scratch.

When planning a display, keep in mind the three Basic Principles of Displaying discussed earlier: (1) the display area should be dark; (2) the background of the display should be dark-colored, nonfluorescent, and nonreflective; and (3) the ultraviolet lamp should be positioned for maximum effect. All three principles apply to all displays of fluorescent minerals, whether large or small, casual or formal, public or private.

Displays With Hand-Held Ultraviolet Lamps

Because hand-held ultraviolet lamps can be moved and brought close to one specimen after another, displays using them can assume almost any shape and occupy almost any space. Most typically, these informal displays are laid out on bookcase shelves. If the shelves are deep and the specimens relatively small, steps can be constructed so that two or more ranks of specimens appear on each shelf. These steps can be made from lumber, styrofoam insulation, cardboard boxes, or whatever else is on hand. If the step material is fluorescent or conspicuously irregular, it can be painted flat black or covered with dark-colored nonfluorescent fabric to mask its imperfections. Although daylight appearance counts, the way a display looks under the ultraviolet lamp is of paramount importance.

Informal bookcase or shelf displays can be small or take up entire walls. In restricted living areas, wall displays often constitute the most efficient use of space. If the specimens are on open shelves, collectors will often protect them from dust and direct light with curtains, plastic sheets, or panel doors. Other collectors prefer to use glass-front cabinets, notably the stacking bookcase sections with glass doors which can be raised to show the treasures within.

Collections can be informally displayed even if they are kept in shallow-drawered mineral cabinets or cardboard boxes, simply by opening the individual drawers or boxes. As long as minerals thus displayed are not wrapped in fluorescent paper or sitting on a fluorescent label on fluorescent cotton in a fluorescent box, all will be well. In a pinch, you can take specimens out of boxes or drawers and display them to good advantage on any nonfluorescent, nonreflective surface, such as a kitchen table with a black cloth thrown over it. As long as the fluorescence of your minerals is reasonably obvious, deficiencies in formal "showmanship" can generally be made up through enthusiasm.

Displays With Fixed Ultraviolet Lamps

Display lamps, which are generally larger, heavier, and more powerful than hand-held ultraviolet lamps, dictate the terms of the displays they are built into. Obviously such displays must be built so that the specimens in them are within the effective range of the lamp. Because this range is determined by the condition of the lamp's filters and tubes, and by the fluorescent brightness of the minerals on display, the overall size of such displays can vary. Still, some general rules apply to their overall design; most are boxes with steps inside to support the specimens and keep them at a more-or-less even distance from the lamp or lamps. Displays with fixed lamps also generally have fronts of glass or plastic. These protect the specimens from dust and sticky fingers, and in a shortwave display the casefront also shields the viewers from stray shortwave ultraviolet light.

On the whole, constructed displays of fluorescent minerals will have either shortwave or longwave ultraviolet lamps, but rarely both. The qualities of fluorescence produced by either type of lamp differ; minerals in shortwave displays generally fluoresce in bright, sharp colors, while longwave displays are noted for muted, pastel colors. Shortwave and longwave lamps are also likely to differ in the area of display each can illuminate, or the distance each should be from the minerals displayed.

Shortwave Ultraviolet Display Lamps

Most shortwave ultraviolet display lamps are built around one or two 15-watt germicidal tubes. Assuming that the tubes and filters are in good condition and that your minerals fluoresce brightly, models with one tube can be used at a distance of about 18 inches and will cover an area of about 18 inches by 30 inches. A two-tube (30-watt) lamp (*fig. 2*) will provide about the same intensity at a distance of 24 inches over an area of 24 inches by 40 inches or more. Larger and more powerful display lamps are made, including lamps with two 30-watt tubes. (These 60-watt display lamps are almost twice the length of the 30-watt lamps, and as you might expect, cover an area almost twice as long and somewhat wider.) The most powerful display lamps in current production have grids of fused quartz tubing. With the same filter area as a typical 30-watt display lamp, they can cover an area of about twice the size at nearly twice the distance. In general, long 60-watt lamps and the quartz grid lamps are for museum displays. For example, the Franklin Mineral Museum in Franklin, N.J. has a display four feet high and deep by 32 feet long, powered by seven quartz grid lamps at a distance of two feet or more from the specimens *(fig. 3)*.

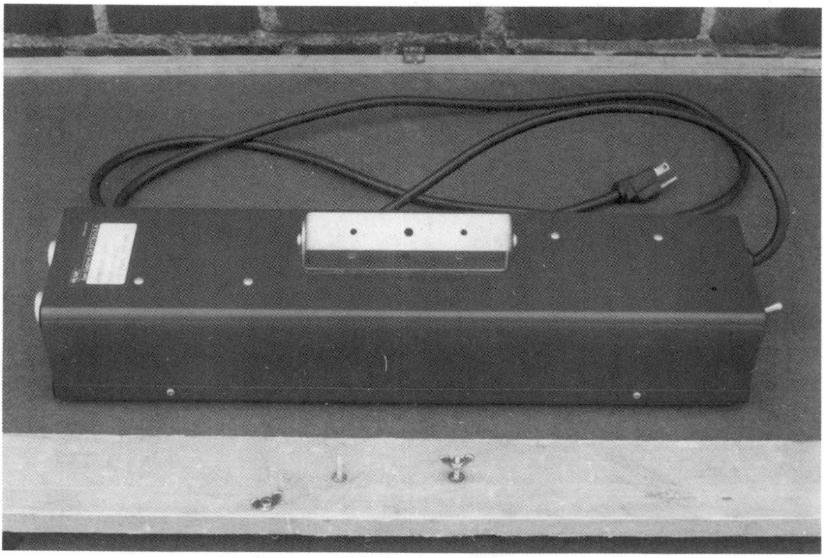

Figure 2. Typical 30-watt shortwave ultraviolet display lamp showing mounting bracket. In foreground, 1 in. x 3 in. x 4 ft. brace with mounting bolts and nuts.

Most collectors do not own museums and do not require the more powerful, more expensive lamps. Excellent home displays and exhibits for mineral shows can be prepared with 15-watt and 30-watt shortwave units or 30-watt longwave units.

Longwave Ultraviolet Display Lamps

Longwave ultraviolet lamps fall into three types. The first, which is also the best for displaying minerals, is that which uses a longwave tube and a separate filter; the tube appears similar to those which produce visible light, in that it is made of clear glass with its inner surface coated. The second type uses dark blue-violet tubes, called "black light blue" or BLB tubes, without a separate filter. The third type uses floodlight-style bulbs with separate filters which also act as lenses; these lamps are often called "bullet lamps" because of their shape.

The first of these three types is configured much the same as most shortwave ultraviolet lamps. The typical display model

Figure 3. The fluorescent mineral display at the Franklin Mineral Museum, Franklin, N.J.

has two 15-watt tubes, and differs in appearance from a standard 30-watt shortwave lamp chiefly in having a larger filter. This is the type of longwave lamp recommended for displays.

The second type of longwave lamp, with BLB tubes, is much more widely distributed than the first. BLB tubes are available in all standard fluorescent tube sizes and will fit standard fixtures. These tubes do not require separate filters because their glass envelope is made of filter glass. However, BLB tubes pass fairly high levels of visible light, making them relatively ineffective for mineral displays. BLB tubes are widely used in theaters, nightclubs, etc., and are certainly the least expensive source of longwave ultraviolet light.

The "bullet" lamp, the third type of longwave lamp, is unquestionably powerful but difficult to incorporate in a mineral display. These lamps illuminate a rather small area for their size; furthermore, they take several minutes to warm up, and once turned off can be difficult to start again.

Adapting the "Federation Case"

Before you build your own display cabinet for a mineral show, you should know that conveniently sized cabinets already exist and that they are easily adapted for fluorescent minerals. Generally called "Federation Cases," after the pattern specified by the American Federation of Mineralogical Societies, they are in use at most mineral shows. They are usually two feet deep, two feet high, and four feet long, with a glass front. For daylight displays they are often illuminated by two floodlight bulbs fastened to the top of the case toward its front. It is generally possible to adapt "Federation Cases" for fluorescent mineral displays simply by removing the floodlight brackets (which are on the outside of the case) and replacing them with an external brace for a single 30-watt ultraviolet lamp, which is mounted inside the case. (Two 15-watt display lamps mounted end to end will do as well.) At the annual Franklin-Sterling Gem and Mineral Show, this is done by placing a plank four feet long and 1x3 inches wide on top of the case, and drilling two 9⁄64-inch holes through the plank and the top of the case to match the spacing of the holes in the mounting bracket of the lamp. Then two 1½-inch long x ⅛-inch bolts (*figs. 2 and 4*) are pushed through the support plank into the case top, and the display lamp is attached with wing nuts. The only modification to the case is two small drill holes. These are placed about three inches from the front edge of the case so that the lamp's energy is directed away from the glass case front but the lamp itself interferes as little as possible with the viewer's line of sight to the specimens (*fig. 5*). While there are undoubtedly many other possible case-and-lamp configurations, this one is recommended for its simplicity and because of the wide availability of this type of case. Display lamps can also be mounted with chains.

"Federation Case" liners are generally constructed of light-

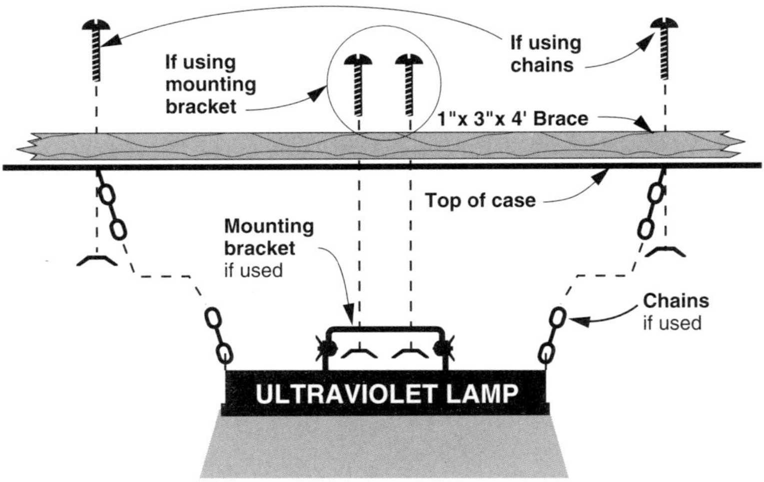

MOUNTING AN ULTRAVIOLET LAMP
Using either mounting bracket or chains
FIGURE 4

weight materials and are often designed to be assembled and torn down quickly. The inside back and sides of the case are usually lined with panels of Masonite or thin plywood covered with cloth. These frame a set of risers, usually constructed of wood 2 x 4s or styrofoam and covered with the same cloth as the back and side panels. (The risers in fig. 6 are easily made from readily available materials but are only suggestions; many other designs are possible.) If part of the floor of the case is exposed, it can be covered with cloth or another thin cloth-covered panel. Once a case like this has been set up, it can be used with equal success at home or in a succession of mineral shows.

Glass and Plastic Case-Fronts

Either glass or plastic case-fronts will block shortwave ultraviolet radiation. Plastic has the advantages of being lightweight and break-resistant but is easily scratched; because of

Figure 5. Typical 2 x 2 x 4 ft. "Federation Case" (without front) adapted for a fluorescent mineral display, with 30-watt shortwave ultraviolet display lamp bolted to external 1 in. x 3 in. x 4 ft. brace. Side and rear panels of ⅛ in. Masonite sheet; panels, styrofoam risers, and case floor covered with nonfluorescent, nonreflective cloth.

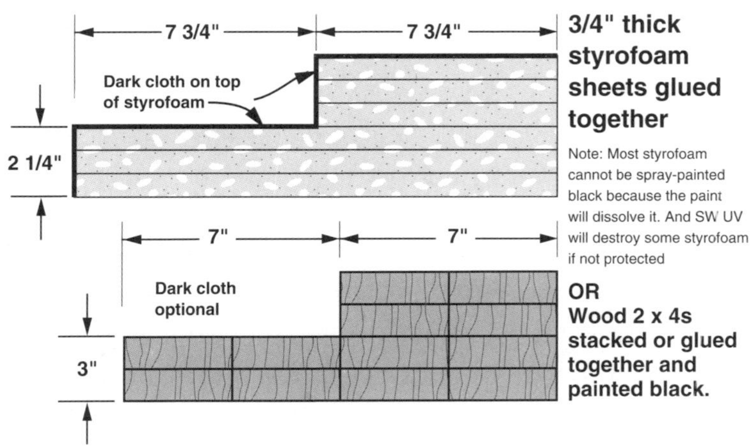

CROSS-SECTIONS OF TYPICAL RISERS

Using styrofoam or wood 2 x 4s

FIGURE 6

this it will probably need replacement eventually. Glass is harder and can be cleaned repeatedly without scratching, but is relatively fragile and heavy. When selecting a case-front material for a display of fluorescent minerals, check both plastic and glass for fluorescence. Almost all glass fluoresces on one side, and in a case-front the fluorescent side should face out, away from the specimens. Some plastic glazing, too, is fluorescent. Whether made of glass or plastic a fluorescing case-front should be avoided; it can ruin a display.

Display Cabinets for Brightly Lit Areas

It is taken for granted that most fluorescent mineral displays will be viewed in a darkened room. If such a room is not available, cases with viewing ports can be obtained or built. The most efficient have apertures configured like goggles to fit the face and exclude daylight. If one is willing to wait with one's face pressed against the port until the eyes adapt somewhat to the dark, the fluorescence of the minerals within can be seen. Another approach to the lit-room problem is a case constructed with mirrors. Here light is more or less excluded from the specimen area by baffles. An ultraviolet lamp is located next to the specimens, and their fluorescence is reflected to a shrouded viewing slot by mirrors.

Both the viewport and mirror/baffle cases allow the viewing of fluorescent minerals in a lit room. Neither type allows the viewer to see many specimens at once, nor are the viewer's eyes likely to be really adapted to darkness so that the full intensity and brilliance of fluorescent minerals can be experienced. However, such cases are perfectly suited for mineral dealers who must exhibit their wares under normal light, and portable cases can be useful in the field when collecting by day.

For schoolroom or science fair use, it is possible to build a simple viewing cabinet from a cardboard box. Two slots are cut in the top of the box, one for the ultraviolet lamp and the other to look through. Specimens can be affixed to the bottom of the box along with their labels, or the bottom can be left out entirely and the box lifted and placed over the specimens.

Custom Display Cabinets for the Home

Permanent home displays are generally in the form of box-like cabinets supported several feet off the floor. Some collectors enclose the bottom section and store minerals there, while others leave it open. The display section of the cabinet typically measures from three to four feet in height and depth, with a front opening anywhere from 2½ to 4 feet high and from 4 to 10 feet wide. Inside the cabinet is a set of steplike terraces rising from front to rear and proportioned to the size of the specimens. These terraces are made of wood planks supported on frames (like a flight of stairs) and are painted flat black. In a small case there will be one display lamp mounted top-center at the front of the case. In a larger case there may be two or more lamps spaced evenly along the top front. In an unusually large display, supplemental smaller lamps may be added at either end, mounted vertically on the side frames.

To protect the viewer's eyes from shortwave ultrav et energy and the specimens from dust and theft, the fron f these displays are usually covered with glass or plastic. ss fronts consist either of plate glass permanently framed lace, or of two plates mounted in a sliding frame. In the r, access to the collection is through those sliding plate g doors. In displays with fixed-glass fronts, access is thro doors on one or both sides of the cabinet. In very large ays of this type *(fig. 3)* there may be a walk-through co between the glass front and the specimens. Rigid plast nts are mounted in wooden or aluminum frames. Thei weight allows them

Sterling Hill, New Jersey. Groupings of these can be exhibited to advantage in a display case, but here a timer is necessary. These timers, or their components, are available through electronics supply stores and catalogs, or from specialized mineral dealers with an interest in fluorescence.

A typical timer sequence for demonstrating mineral phosphorescence would be to turn the ultraviolet lamp on for ten seconds and off for five seconds, over and over. More complex displays of phosphorescence can include an interval where the daylight appearance of the minerals is shown through inclusion of low-level (7- or 10-watt) incandescent bulbs in the case. Here the timer sequence might be 10 seconds of shortwave ultraviolet light, 5 seconds of darkness, and 7 seconds of artificial light. Obviously a timer built for the latter sequence would have to have two electrical outlets, one for the ultraviolet lamp and the other for the incandescent bulbs.

There are dramatic ways of demonstrating the duration of mineral phosphorescence and its response to heat. The latter can be done simply by plunging a phosphorescing mineral into a glass beaker of hot water, or by holding that mineral under the flow from a hot water tap. A specimen of phosphorescing white willemite from Franklin, New Jersey, for example, will flare up brightly when this is done; the effect can be repeated several times in succession before the visible phosphorescence is exhausted.

The duration of phosphorescence varies enormously from mineral to mineral but can be shown vividly with a sphere of red-fluorescing calcite such as that from Franklin, New Jersey. This calcite has a brief orange-red phosphorescence. If a shortwave ultraviolet lamp is directed at one side of a spinning sphere of this calcite, the phosphorescence will be seen to carry over to the opposite side of the sphere. If the spinning slows down, the area of phosphorescence will shrink; if it speeds up, the phosphorescing area will grow. In a live display, the sphere can be spun by hand. At least one collector has built a display with a calcite sphere spun by an electric motor.

Displaying Tenebrescence

A rare but extraordinary feature of a few fluorescent minerals, notably sodalite and tugtupite, is the property of changing color in daylight on exposure to shortwave ultraviolet energy. This phenomenon is called tenebrescence or reversible photosensitivity. The hackmanite variety of sodalite exhibits this property to a remarkable degree; good specimens can change from white to deep reddish purple in a minute or less of exposure to shortwave ultraviolet light. Bleaching of the reddish purple color under incandescent light or daylight can be equally rapid. Because of the time involved, this remarkable change is best demonstrated in a live display where the exhibitor can talk about something else while the color transformation is underway.

Displaying Triboluminescence

Triboluminescence is light emitted when certain minerals are struck, scratched or rubbed. This can be shown only in a live display in a very dark room, as the light level of triboluminescence generally is quite low. Such demonstrations are obviously not recommended for fragile or valuable minerals. Massive quartz is frequently triboluminescent. Sphalerite from some localities is so sensitive that it will glow when scratched by a fingernail or a piece of stiff plastic. Triboluminescence will often be seen in mines or on nighttime field trips if it is looked for when rocks are struck together, hammered, or drilled.

simple viewing cabinet from a cardboard box. Two slots are cut in the top of the box, one for the ultraviolet lamp and the other to look through. Specimens can be affixed to the bottom of the box along with their labels, or the bottom can be left out entirely and the box lifted and placed over the specimens.

Custom Display Cabinets for the Home

Permanent home displays are generally in the form of box-like cabinets supported several feet off the floor. Some collectors enclose the bottom section and store minerals there, while others leave it open. The display section of the cabinet typically measures from three to four feet in height and depth, with a front opening anywhere from 2½ to 4 feet high and from 4 to 10 feet wide. Inside the cabinet is a set of steplike terraces rising from front to rear and proportioned to the size of the specimens. These terraces are made of wood planks supported on frames (like a flight of stairs) and are painted flat black. In a small case there will be one display lamp mounted top-center at the front of the case. In a larger case there may be two or more lamps spaced evenly along the top front. In an unusually large display, supplemental smaller lamps may be added at either end, mounted vertically on the side frames.

To protect the viewer's eyes from shortwave ultraviolet energy and the specimens from dust and theft, the fronts of these displays are usually covered with glass or plastic. Glass fronts consist either of plate glass permanently framed in place, or of two plates mounted in a sliding frame. In the latter, access to the collection is through those sliding plate glass doors. In displays with fixed-glass fronts, access is through doors on one or both sides of the cabinet. In very large displays of this type *(fig. 3)* there may be a walk-through corridor between the glass front and the specimens. Rigid plastic fronts are mounted in wooden or aluminum frames. Their light weight allows them to be hinged so that they can be raised or swung out of the way, or simply dismounted to reach the specimens.

It is possible to construct a so-called permanent home display so that it can be broken down and carried (in whole or part) to a mineral show on those occasions when a large display is required. However, this usually demands a good deal of ingenuity in the design of the case and considerable time and muscle to move it, set it up, and tear it down again.

Special Fluorescent Mineral Displays

Unusual and exciting displays can be put together around some of the distinctive properties which some minerals possess. These include A) *phosphorescence,* where a mineral continues to emit light after the exciting ultraviolet radiation is switched off; B) *tenebrescence,* in which a mineral changes its daylight color during exposure to ultraviolet energy; C) *triboluminescence,* the light emitted when a mineral is scratched or struck; and D) the ability of certain minerals to fluoresce different colors when the wavelength of the exciting ultraviolet radiation is changed. Other original displays can be developed by the lapidary who cuts and polishes fluorescent minerals or the artist who uses crushed minerals as a medium.

Displaying Phosphorescence

Phosphorescence is the easiest of these properties to demonstrate. Many species exhibit it, often conspicuously. If the display is "live," with the specimens handled by their owner, little preparation is needed to expose a mineral to ultraviolet energy, switch off the ultraviolet lamp, and show the phosphorescing mineral to the audience.

Calcite and aragonite from worldwide localities often show conspicuous phosphorescence; so do rarities like tugtupite, and the well-known phosphorescent willemite from Franklin and

Displays Incorporating Shortwave and Longwave Lamps

Some minerals fluoresce different colors when exposed to different wavelengths of ultraviolet light. The best-known is calcite from Terlingua, Texas *(plates 2-4)*; under shortwave ultraviolet light it fluoresces and phosphoresces bright medium blue, yet when irradiated with longwave ultraviolet light, it fluoresces a distinct pink color. Some aragonite specimens from Sicily have an even more complicated response: they fluoresce white under shortwave ultraviolet light and pink under longwave ultraviolet light, but phosphoresce green after exposure to light of either wavelength. It is possible to show these fluorescent color changes in a display with a timer similar to that used to exhibit phosphorescence; first the shortwave ultraviolet lamp is turned on by itself, then the longwave ultraviolet lamp by itself, then both are turned off.

When both shortwave and longwave ultraviolet lamps will be used in a display it is possible to combine their effects, often with surprising results. There are many specimens which contain more than one fluorescent mineral, and not infrequently these minerals will respond to ultraviolet light of different wavelengths. One example is certain sphalerite specimens from the Sterling mine, Sterling Hill, Ogdensburg, N.J. The sphalerite fluoresces orange with blue highlights under longwave ultraviolet light, while the associated willemite and calcite fluoresce green and red, respectively, under the shortwave lamp. Seen under shortwave and longwave lamps simultaneously, these specimens are exceptionally attractive.

Combined Displays of Distinctive Properties of Fluorescent Minerals

An exhibit which combines many of these properties

(phosphorescence, minerals which fluoresce different colors under shortwave and longwave ultraviolet light, and specimens with several minerals whose fluorescence can be seen together only under combined shortwave and longwave ultraviolet light) has been created by Don Newsome, founder of the Fluorescent Mineral Society. His display contains shortwave and longwave ultraviolet lamps, low-wattage incandescent bulbs, and a timer which is activated when a spectator steps on a mat in front of the display case. Prior to activation, the longwave lamp is on steadily. When the timer cycle begins, the longwave lamp is switched off and the shortwave lamp on for 10-12 seconds; then both lamps are on together for another 10-12 seconds, followed by a 7-8 second period of darkness to show phosphorescence. Following this, the incandescent bulbs are brightened and dimmed over about 15 seconds to show how the minerals appear in daylight; then the display returns to its static state with the longwave lamp on, ready for the next timer cycle. During this sequence a tape recorder explains what is being seen. While not every collector will want to build such a display, it suggests what can be done with a timer, shortwave and longwave lamps, and some ingenuity.

Arts and Crafts with Fluorescent Minerals

The craft of the lapidary can include the use of fluorescent minerals as a medium. Quite a few collectors are addicted to fluorescent spheres, eggs, slabs, or cabochons, and in truth a display of any of these can be impressive. Often the techniques and equipment necessary to cut and polish fluorescent minerals are within the reach of the average collector, and it is not unusual for many mineral enthusiasts to involve themselves in more than one aspect of their hobby. In all fairness, it must be mentioned that there is a widespread prejudice against destroying a fine specimen to make a pretty sphere or belt buckle; I share this prejudice. However, some fluorescent minerals are abundant and exist in large masses of no overwhelming aesthetic or scientific interest. These may as well be sphered or slabbed as broken up for hand specimens. Rarely one also hears of fluorescent coffee tables, fireplaces, rock gardens, and even religious shrines! In their crafted or constructed form these fluorescent minerals will often attract the attention of people who might have ignored the same minerals in their natural state.

The ultimate in destruction of specimens to further a craft is the pulverizing of fluorescent minerals in order to paint pictures with them *(plate 15)*. The technique is very much of the paint-by-numbers school. An outline is made on plywood, masonite, or other stiff backing, and the coloring planned in discrete areas, with each area designated for a particular mineral or mixture of minerals. A nonfluorescent glue (typically a white, water-based glue) is applied to an area, and the crushed mineral sprinkled on that glue, which is then allowed to dry; generally one mineral at a time is added to the picture. The results vary considerably, depending on the talent and imagination of the artist. The possible effects are somewhat limited, as are the number of fluorescent minerals fit for the artist's palette, but the effort may well be worthwhile, and the technique certainly has not been exploited to its furthest reaches.

Lecturing About Fluorescent Minerals

Giving public talks about fluorescent minerals is a logical extension of displaying them. Both are part of sharing your knowledge and enthusiasm. Because the appeal of fluorescent minerals is so wide, you will find responsive audiences not only at mineral clubs, but also at schools, nature museums, civic organizations, scout groups, etc. At mineral shows where there is a fluorescent display, there may also be interest in an accompanying lecture which explains fluorescence.

With time you may expand your lecture paraphernalia to include such items as a sphere for demonstrating phosphorescence; drawings, paintings, or slides illustrating properties and applications of fluorescence; household items which fluoresce; slides of fluorescing minerals; and, in fact, anything else which serves to illustrate your concerns and which will enlarge the understanding of your audience.

Some basic rules of showmanship apply to lectures as much as displays: have good ultraviolet lamps and brightly fluorescing minerals, and make sure the lecture room is really dark! Your talk, like decent display labels, should be clear, concise, and to the point. In the beginning, it is a good idea to develop and practice your lecture before family and friends. Their suggestions will be helpful, even when couched in negative terms. ("Don't stand with your back to the audience." "Don't mumble." "Don't say 'Um' or 'You know' in every sentence.") When you finally speak in public, you will be relieved to find that you and your audience have something in common; you want to inform and entertain, and the audience wants to be informed and entertained! The experience is nearly always positive for all concerned.

A typical introductory lecture about fluorescent minerals might have an outline like this:

1) Describe ultraviolet light and its relationship to visible light and other forms of energy. Distinguish shortwave from longwave ultraviolet light.
2) Explain how an ultraviolet lamp works.
3) Define fluorescence. Explain what happens inside atoms when fluorescence takes place.
4) Give some examples of fluorescence in everyday life: TV screen phosphors, "day-glow" colors, fluores-

cent lighting, postage-stamp "tagging," etc.

5) Describe how prospectors looked for tungsten and uranium ores with the ultraviolet lamp. Illustrate with samples.

6) Display examples of the most spectacular fluorescent minerals, which are collectors' favorites just for their beauty.

7) Demonstrate some unusual features of fluorescent minerals, such as phosphorescence, tenebrescence, minerals which fluoresce different colors under different wavelengths of ultraviolet light, etc.

8) Focus on local mineral collecting. Display local minerals, mention local mineral clubs, museums, and shows.

9) Remind the audience of the importance of personal safety in the use of ultraviolet lamps and of the value of conserving both specimens and mineral localities.

10) Conclude, emphasizing the beauty of fluorescent minerals and the joy of collecting them.

Obviously such talks have to be geared to the audience, and you cannot expect a group of Cub Scouts or Brownies to react the same way as a high school science class or the members of a Rotary Club. Small children enjoy colors and will want to touch your specimens, but will lose interest if you deviate into technical explanations. High school science students might appreciate a brief technical explanation of fluorescence, and should be reminded of the practical applications of fluorescence and ultraviolet radiation in modern technology.

As you develop your own collecting interests and specialties you will want to include them in your talks; many collectors are capable of lecturing on multiple subjects. Your knowledge is important, but personal details and anecdotes can also be woven into the fabric of a talk to give it more depth and color. A good lecture about fluorescent minerals will excite, entertain, and inform; it will also reward the speaker as well as the audience.

Specialized Collecting

In an earlier part of this chapter much was said about why collectors collect, and how they can acquire, maintain, and display their collections. Little was said about what to collect, or why.

Collections of fluorescent minerals evolve along with their collectors. Because no two people are the same, all such collections will be different and will reflect the unique tastes and interests of their owners. Ultimately every collector specializes in something, and these specialty areas often become the most interesting and significant parts of a collection. Here is where intermediate and advanced collectors can learn the most and be in a position to pass this knowledge on, personally and through their specimens and even their writings.

Almost every collector of fluorescent minerals begins in the same way, by acquiring the brightly fluorescing, widely available classics such as scheelite; hydrozincite; hyalite opal; fluorite; Canadian "wernerite;" the blue/pink-fluorescing calcite from Terlingua, Texas; willemite with calcite from Franklin, New Jersey; and others. Part of the definition of a fluorescent mineral classic includes the fact that every collector knows about and wants them; these classics seem to be described and photographed in almost every book about fluorescent minerals, and most general displays will feature several of them at least. Acquiring them is like learning the basic vocabulary for a new language; it is necessary before you can go further.

Specializing is inevitable; as you collect, certain things will interest you more than others. The more closely you pursue those interests to the exclusion of others, the more your collection will develop depth, and the more you will discover about your specialty. Looked at another way, fluorescent minerals come in an immense variety of shapes, sizes, colors, prices, and species from thousands of localities all over the world, and you CAN'T collect them all. Better to concentrate on what fascinates you most.

As your collecting evolves, you may discover that you have formed several collections, each with a different focus. These specialized collections will interlock as part of a greater whole. Whatever your collecting goals, and whether your collection is small and tidy or enormous and sprawling, your activities will be a source of endless fascination and delight. By following your interests wherever they lead, you cannot go wrong. Here are some examples of specialty collecting areas:

- Specimens with two or more fluorescent minerals.
- Fluorescent mineral spheres or gemstones.
- Phosphorescent minerals.
- Crystals of fluorescent minerals.
- Triboluminescent minerals.
- Fluorescent ore minerals.
- Rare fluorescent species which are known in *good* specimens from one locality only, e.g., tugtupite, benitoite, esperite.
- Minerals which fluoresce more than one color in a single specimen. Some minerals (e.g., sphalerite) are capable of fluorescing more than one color under one wavelength of ultraviolet light. Others will fluoresce different colors under longwave and shortwave ultraviolet light.
- All fluorescent varieties of a particular mineral

species, group, or chemical class, e.g., calcite, fluorite, willemite; amphiboles; halides, carbonates.

• All fluorescent minerals from a particular locality, e.g., Franklin and Sterling Hill, New Jersey; Långban, Sweden; Tsumeb, Namibia; Mont St. Hilaire, Quebec; the state of Iowa; the nation of Israel; etc.

The last two specialties above, which are attempts to understand the diversity of fluorescent minerals within chemical and physical bounds on one hand, and geographical bounds on the other, are potentially the most complex and rewarding of these specialties. They can lead not only to years of enjoyable collecting but also to more permanent accomplishments.

Every collector with one or more specialties gradually acquires a depth of knowledge in those areas which can be passed on through displays and talks, as previously described, and ultimately through writing and publication. Mineral-related publications range from informal club newsletters to formal, highly technical journals, with many collector-oriented journals in between. Depending on your experience and interests, you may be in a position to contribute useful knowledge at any of these levels. There is a widely accepted aphorism in the scientific community that if knowledge isn't written down, it doesn't exist. Practically, your words on paper exist independently of you, and through publications you will inform and influence many more people than you could meet in a lifetime.

For your collection to grow and your interests to develop, you need help. It will come from the community of mineral collectors, past and present: from those who formed collections and gave them to museums so that you could study them; from others who wrote down what they knew so that you could read it; from fellow collectors who will share their time and knowledge and, sometimes, their minerals; and from dealers who can help you build your collection. You will also have many opportunities to give something back to that community. Besides displaying your minerals, lecturing about them, and perhaps writing about them, you can also participate in mineral clubs or civic organizations and work with local institutions. Help to organize, set up, and tear down a mineral show; take a scout troop on a field trip; prepare a collection of minerals for donation to a local school. Be a giver, not just a taker. And remember that as your knowledge increases, so does your responsibility to use it for the benefit of others.

Franklin and Sterling Hill, New Jersey: More Than Just Another Specialty

In collecting, we may be drawn to the remote and exotic, or to the near and familiar. In my case, the Franklin-Sterling Hill area is all of these. I grew up 25 miles away, but when at age 17 I first became aware of the Franklin area's minerals, their oddity and beauty made me think of treasure from Aladdin's cave. Over 30 years later, the impression has not dimmed. To be sure, collecting the fluorescent minerals of Franklin and Sterling Hill is a specialty, but pursuing it has involved nearly every aspect of mineral collecting mentioned in this chapter.

The Franklin-Sterling Hill district is incontestably one of the world's very great mineral localities and would be so even without the distinction of its fluorescent minerals. It has the world's highest total of mineral species from one locality, over 340 as this is being written (1994), and world records both for the number of minerals found nowhere else and for the number of minerals first described from a locality. Even its ores are extraordinary. Franklinite (zinc iron manganese oxide) and zincite (zinc oxide), both present in huge quantities locally, are known elsewhere only in trace amounts. Willemite (zinc silicate), the second most abundant of the

Franklin-Sterling Hill zinc ores, is found sparsely throughout the world as a weathering product of sphalerite and is rarely mined; in the Franklin area it is a primary mineral which also happens to have a variety of colors and forms unequaled by any other mineral from any other locality. Most specimens from the Franklin-Sterling Hill district are so distinctive in appearance that almost no-one could confuse them with mineral samples from another locality.

There were two major zinc orebodies in the district, both with the same basic mineral content and geologic history. The larger of the two was in Franklin. There the ore was considered richer and easier to mine than at the companion orebody at Sterling Hill, two-and-one-half miles to the southwest and adjacent to the town of Ogdensburg. By 1954, when the Franklin mine closed, over 22,000,000 tons of the world's richest zinc ore (typically close to 20% zinc) had been removed. The Sterling mine at Sterling Hill yielded about half as much ore as the mine in Franklin before closing in 1986.

Both orebodies had been known to science since the beginning of the nineteenth century, and the volume of scientific literature relating to them is enormous: over 1000 papers, reports, monographs, and the like as of 1994. Exploitation of the ores had been attempted before 1800, when small iron mining and smelting operations were widespread in the New Jersey Highlands, but full-scale mining was delayed until well after the Civil War. The reasons were both technical (the ores were difficult to smelt with existing techniques) and legal (the rights to mine iron and zinc from the same orebodies had been sold separately, even though the major ore mineral, franklinite, was an ore of both). In its day, the Franklin-Sterling Hill district was one of the greatest zinc-producing areas in the United States; geologically and mineralogically, it is still not perfectly understood, and in the scientific sense its history is far from over.

The Franklin-Sterling Hill district encompasses not only the two zinc orebodies but also the local portion of their host formation, the Franklin Marble. Several marble quarries in the district have contributed many fluorescent minerals to the current total, some of them (e.g., norbergite and diopside) worthy of inclusion in any display.

To most collectors of fluorescent minerals, "Franklin" is typified by the red-and-green-fluorescing combination of calcite and willemite. While these two are found together at a dozen or more other localities, at none of them is this combination abundant or typical. At Franklin and Sterling Hill it was a substantial proportion of the millions of tons of ore mined there. Many thousands of pounds must have escaped into the hands, car trunks, and basements of thousands of collectors. This willemite-calcite combination is the most widely distributed and appreciated fluorescent specimen material in the world *(plate 30)*.

Nearly one out of four Franklin-Sterling Hill minerals fluoresces, a total (82 in 1994) which again is a world record. Of these, over 20 are easily displayable and more than ten among the brightest fluorescent minerals found anywhere. Beyond this, what is truly exceptional about Franklin and Sterling Hill is that two or more of these minerals are often found together in visually striking patterns. Franklin-Sterling Hill ore is a high-grade metamorphic rock with a long (over one billion years) and complicated history; naturally the textures of the ores reflect this. Besides highly variable grain size, there are folds, faults, veins, breccias, layering, and other evidences of rock formation, distortion, and movement. All these can be seen in hand specimens. When these patterns are illuminated in bright red, green, yellow, orange, violet, and blue, the overall effects are staggering.

To the collector of fluorescent minerals there are some important differences between Franklin and Sterling Hill. Of the brightest fluorescent minerals from the district, six are not found at Sterling Hill: esperite *(plates 20, 26)*, clinohedrite *(plates 17, 24)*, hardystonite *(plates 17, 26)*, manganaxinite[2], margarosanite *(plate 24)*, and turneaureite *(plate 19)*. When the Franklin mine closed in 1954 the supply of these was eliminated, and the demand has since been supplied from miners' accumulations, dealer stocks, and the recycling of collections. Sterling Hill has continued until recently to produce superb fluorescent willemite, calcite, sphalerite *(plate 23)*, wollastonite *(plate 29)*, hydrozincite, barite, fluorapatite, and johnbaumite. Sterling Hill and the nearby Franklin (Farber) quarry have also recently yielded excellent specimens of the fluorescent minerals from the Franklin Marble: norbergite, tremolite, diopside *(plate 27)*, fluoborite, and others.

Although both the Franklin and Sterling mines are now closed and flooded, the district still has many attractions for the mineral collector. Of the two museums there, the Franklin Mineral Museum has extensive collections of local and worldwide minerals and a gigantic (32 feet long) fluorescent mineral display *(fig. 3)*; the Sterling Hill Mining Museum, on the site of the Sterling mine, exhibits an abundance of mining equipment and features an underground tour which includes "The Rainbow Room," a striking show of fluorescent minerals incorporating zinc ore *in place*. Both museums are open from March through November, and both have mineral dumps where you can collect for a fee. At any time of year local mineral dealers can be found operating out of their car trunks or

---

[2] Manganaxinite is found in small amounts at Sterling Hill but is not known to fluoresce from that locality.

basements; however, the best time to visit the Franklin-Sterling Hill area is in early Fall. This is when the Franklin-Sterling Gem and Mineral Show, an annual event since 1956, combines the above attractions with informative lectures, mineral exhibits, and an abundance of mineral dealers. If there is a single mandatory event for collectors of fluorescent minerals, this is it.

In 1968 the New Jersey Legislature designated Franklin "The Fluorescent Mineral Capital of the World." This is not political hype. Franklin and Sterling Hill, with their geological and mineralogical complexity and lengthy scientific and cultural history, comprise in any case a mineralogical Mecca. The astonishing variety of fluorescent mineral species, associations, and patterns found there has made the Franklin-Sterling Hill district *the* ultimate locality for collectors of fluorescent minerals. After more than 30 years of collecting the local minerals, I can truthfully say that their fascination is endless. Every Franklin-Sterling Hill collection I look through has something I have never seen before: old minerals in new settings; unexplained fluorescences and unexpected fluorescent juxtapositions; the familiar and the enigmatic in near-infinite succession. For me and many others, the Franklin-Sterling Hill district is the greatest mineral locality in the world.

Postscript

Perhaps the true wonder of collecting is that the world contains so much to fascinate us. If just Franklin and Sterling Hill are so rich that one cannot collect or completely understand one aspect of them in a lifetime, how splendidly abundant in complexity must the rest of the world be? Those of us who have been drawn to fluorescent minerals—one strand in the infinite web of nature—must hope that in enjoying these minerals, conserving them, learning about them, and educating others about them, we are all doing our part to appreciate and preserve one more unique aspect of our extraordinary planet.

Acknowledgements

In preparing this chapter I have been aided by comments from Tema Hecht, Don Newsome, and Thomas Warren. The bulk of helpful criticism came from Earl Verbeek, who edited and proofread the text at every stage of its development.

Much in this chapter has been derived from the lore and customs of the fluorescent-mineral-collecting subculture which has evolved over the last five decades. Of this culture's adepts who wrote down what they knew, I owe the most to Sterling Gleason, Robert W. Jones, Jr., and Manuel Robbins. I have also learned much from fellow members of the Franklin-Ogdensburg Mineralogical Society, Inc., who have for many years contributed to the displays of fluorescent minerals at the Franklin-Sterling Gem and Mineral Show. In my collecting and writing I have been most strongly influenced by Ethel Packard "Sunny" Cook, Thomas Warren, and Pete J. Dunn.

Special thanks are also due to Virginia Hiramatsu and Don Nelson, who patiently kept up with the hundreds of changes imposed on the text by its authors.

Activators in Fluorescent Minerals

Earl R. Verbeek

Part 1: The Mechanisms of Fluorescence

Introduction

Fluorescence is a fascinating and beautiful property of many minerals, but a most inconsistent one. Consider, for a moment, the mysteries of fluorite—the mineral after which the property of fluorescence was named more than 140 years ago. Fluorite from the Weardale mining district in England shows a remarkably bright, deep violet-blue fluorescence so appealing that large crystal groups are featured items in many museum displays worldwide. Some of the same displays, however, contain blocky fluorite crystals from Clay Center, Ohio, which glow not violet-blue, but bright pale yellow. Nearby might be a rounded, lumpy mass of fluorite showing deep green fluorescence and labeled Fremont County, Colorado; or perhaps a group of lovely, deep red-fluorescing crystals from the Ojuela mine in Mapimí, Mexico. Violet-blue, pale yellow, green, red, all these colors of fluorescence from the same mineral species—no wonder generations of collectors have found fluorescent minerals so intriguing.

To unlock the mysteries of fluorescence, many investigators over the years analyzed fluorescent minerals for chemical clues to their response to ultraviolet light. Others prepared synthetic minerals in the laboratory in attempts to duplicate the response of their natural counterparts. Still others examined the crystalline structure of various minerals at the atomic and subatomic levels to better understand the physical mechanisms of fluorescence. Gradually a body of evidence accumulated that fluorescence in minerals, though highly variable for some species, is nevertheless systematically related to other properties of minerals—their chemistry and structure—in understandable ways. Among the general findings were these:

(1) Nearly all minerals with a metallic or submetallic luster show no visible fluorescence under ultraviolet light. Copper, gold, and silver are familiar examples among the naturally occurring elements; so too are such other common minerals as pyrite, chalcopyrite, magnetite, hematite, and galena. Those minerals that *do* fluoresce almost uniformly have a nonmetallic (dull, waxy, glassy, greasy, resinous, or adamantine) luster.

(2) Nearly all fluorescent minerals are electrical insulators—that is, they are poor conductors of electricity. A few minerals that are natural semiconductors fluoresce also, but most do not. Minerals that are electrical conductors (the metals) do not fluoresce. This property is related to number (1) above, for a close correlation exists between the electrical properties of minerals and their lusters.

(3) Among the many minerals that fluoresce in some specimens but not others, chemical analyses commonly disclose that the most chemically pure samples are those least likely to fluoresce. Examples include some of the species most sought by collectors of fluorescent minerals—calcite, willemite, wollastonite, and fluorite, to name a few. Like their natural counterparts, the same minerals synthesized in the laboratory to high standards of purity show none of the bold fluorescence for

which these species are so well known. Chemical impurity, it seems, is conducive to fluorescence in numerous mineral species.

(4) The above result notwithstanding, minerals that contain appreciable (1% or more) iron, nickel, cobalt, or copper generally show no visible fluorescence, or at best only a feeble glow. A prominent example is siderite, a carbonate of *iron,* for which we know of only a single report claiming visible fluorescence under ultraviolet light. Yet cerussite, a carbonate of *lead,* is known as a fluorescent mineral from dozens of localities, and calcite, a carbonate of *calcium,* is known from many hundreds.

These general observations are consistent with the accumulated experience of hundreds of collectors as they searched worldwide for fine fluorescent material: most mineral species, with some notable exceptions discussed below, show either variable fluorescence or none at all. For many minerals, it seems, the property of fluorescence is an "accidental" one—that is, a property not intrinsic to the mineral itself. Incidentally, the inconstancy of fluorescence helps explain why many mineralogists pay little heed to fluorescence in their studies. Tabulations of mineral properties commonly include such information as chemical composition, specific gravity, refractive indices, cleavage and parting, common crystal forms, and hardness, but they omit mention of fluorescence (and daylight color) for most minerals because these properties are neither constant nor diagnostic of the species. If mineralogists sometimes display a lower level of interest in your glowing treasures than you do, please realize there is good reason. Among scientists the study of fluorescence lies more in the realms of solid-state physics and phosphor chemistry.

What, then, to make of fluorescence? Does the constancy of fluorescence in some species and the variability in others mean that more than one mechanism is involved? Exactly. The agents that cause fluorescence, known collectively as *activa-*

tors, have been the subject of much scrutiny by scientists over the past 60 years. Their research has uncovered an astonishing diversity of activators in minerals, some of them well understood but others not; many require further study. Much of this research was spurred by the growing needs of an evolving society, including (a) the need for improved methods of household and industrial lighting following the electrification of the United States, (b) the demand during World War II for increased supplies of tungsten, the major ore of which is highly fluorescent, (c) expansion of the nuclear-power industry, which during the 1950s and 1960s resulted in much new research on the fluorescence of uranium ore minerals, and (d) the growing need for efficient phosphors for color TV screens, computer monitors, optical lasers, and other equipment. The list could go on and on. Some of what was learned is presented in this book.

Though the discussions that follow are largely nontechnical, some of the concepts introduced below may be new to some readers. Also, the use of a few specialized terms for which no near equivalent exists in common speech cannot be avoided. But to understand minerals and their fluorescence, it is first necessary to become familiar with a few basics of mineral chemistry. The next three sections address some of these topics for those who might not already be familiar with them.

Chemical Symbols

By definition, each mineral species recognized by science is characterized by a definite chemical composition, or a range of compositions within specified limits. Scientists express these compositions by means of symbols that identify the types and proportions of the chemical elements present. The ideal composition of the mineral *sphalerite,* a zinc sulfide, is written ZnS, which tells us that for every atom of zinc (Zn) in the sphalerite structure, there is also one of sulfur (S). Similarly, the formula for *fluorite* is expressed CaF_2, indicating a proportion of one calcium (Ca) atom for every two fluorine (F) atoms. Chemical symbols for those elements important in minerals are listed at the end of this chapter (Table 1).

You may notice immediately that some of the symbols seem not to match their names. Iron, for example, is denoted Fe, lead is Pb, and antimony is Sb. These symbols, and others like them, reflect the ancient names for these substances in some other language, commonly Latin: iron was "ferrum" to the Romans, lead was "plumbum", and antimony was called "stibium". Tungsten, denoted W, owes its chemical symbol to its alternative name in German: Wolfram. The historical development of many branches of the natural sciences is reflected in their widespread use of words and symbols derived from other languages, most notably Latin and Greek; the science of mineralogy is no exception.

Certain combinations of elements are so common in nature that they have been given their own names. The chemical composition of siderite, $FeCO_3$, is expressed in words as *iron carbonate* rather than *iron carbon oxide* because the carbonate group—the CO_3 part of the formula—is common to so many mineral species. The composition of willemite, Zn_2SiO_4, is stated as *zinc silicate* for similar reasons. The literature on mineral fluorescence is filled with reference to arsenate, borate, phosphate, and sulfate minerals, many of which fluoresce. You will find numerous such group names in common use; the most important are listed in Table 2.

Most minerals depart to some measurable degree from their ideal compositions, and in some the departure is marked. Some elements so readily substitute for others in minerals that their common presence in more than negligible amounts is indicated in the chemical formula. Sphalerite is a familiar example; the chemical formula for this mineral often is given as (Zn,Fe)S rather than simply ZnS, the parentheses and comma indicating the common substitution of iron for zinc.

Working with chemical formulas may seem difficult at first, but with practice it soon becomes second nature. Certainly the effort of memorizing symbols for at least the most common elements in minerals is worthwhile. The fluorescent minerals chondrodite and norbergite, for example, are both *magnesium silicate fluoride hydroxides,* but these words fail to convey any difference between them. Their chemical formulas, however, disclose the differing atomic proportions of the elements present: chondrodite is $Mg_5(SiO_4)_2(F,OH)_2$, but norbergite is $Mg_3(SiO_4)(F,OH)_2$. Where appropriate, chemical formulas for minerals are given in the text; a complete list is appended as Table 3.

Ions and Oxidation States

Most of you probably are familiar with the classic picture of an atom: a dense nucleus of protons and neutrons surrounded by a whirling cloud of electrons. The common depiction of electrons in circular orbits around the nucleus is not quite accurate (orbital shapes are actually quite complex), but that need not concern us here. The important point is to realize that the electrons, particularly those in the outermost "orbits," are responsible for many properties of minerals, including daylight color, luster, diaphaneity (whether the mineral is transparent, translucent, or opaque), thermal and electrical conductivity, hardness, and fluorescence. All of these properties are interrelated.

Most of you probably know, too, that protons and electrons are electrically charged, and that an atom of any element contains an equal number of positively charged protons and negatively charged electrons. When atoms combine to form a solid substance, however, the atoms of one element commonly share one or more electrons with the atoms of another element and so form a chemical bond between them; these bonds are what holds the substance together and gives it its hardness. An atom that has lost or gained electrons, and has thereby acquired a net positive or negative electrical charge, is called an *ion*. The amount of charge is called the *valence* or *oxidation state* of the ion and is indicated by such terms as monovalent (1), divalent (2), trivalent (3), tetravalent (4), pentavalent (5), and hexavalent (6). In fluorite, CaF_2, each calcium atom has "donated" one electron to each of two adjacent fluorine atoms; the calcium atom (now an ion) has a net electrical charge of +2 and is said to be in the divalent oxidation state. Similarly, the two fluorine ions are said to be monovalent; each has gained an extra electron and has a charge of -1.

It is important to realize that some elements such as manganese (Mn), iron (Fe), and uranium (U) can exist in more than one oxidation state. Manganese in the mineral species manganosite (MnO) is in the +2 oxidation state but in pyrolusite (MnO_2) is in the +4 oxidation state (oxygen ions are divalent negative). The effect of a given element on the fluorescence of a mineral is strongly dependent on its oxidation state within the crystal structure. Divalent manganese, as discussed later, is a powerful activator of fluorescence in many minerals, but tetravalent manganese is of lesser importance, and then only at low concentrations. When the valence of certain ions is not immediately apparent from the chemical formula, it is sometimes indicated by a superscript, as in manganaxinite, $Ca_2Mn^{2+}Al_2BSi_4O_{15}(OH)$. Here the superscript 2+ indicates that the manganese is in the divalent state.

Major, Minor, and Trace Constituents of Minerals

Mineralogists often speak of major, minor, and trace constituents of minerals. These terms are used in somewhat different ways by different authorities, but one common definition of them is as follows:

Major: any constituent present in amounts of 1% or more.

Minor: any constituent present in amounts of 0.01% to 1%.

Trace: any constituent present in amounts of 0.01% (100 parts per million) or less.

The minor and trace elements often are present as impurities in minerals whose major constituents are expressed in their chemical formulas.

Activators of Fluorescence

Potential difficulties in understanding the cause(s) of fluorescence in any given mineral species are many. Nowhere is this more apparent than in the literature on the fluorescence of sphalerite, ZnS, a common fluorescent mineral and one of the most intensively studied of all laboratory-produced phosphors. Early results from laboratory experiments revealed a bewildering complexity in its response to ultraviolet light. Synthetic sphalerite could be made to fluoresce blue, yellow, orange, red, or green if appropriate amounts of impurities—including silver, aluminum, copper, manganese, tellurium, dysprosium, lithium, and tin—were added to the host material, singly or in combination. Minor differences in composition clearly resulted in major differences in fluorescence. The results also varied with the method of preparation: samples of identical composition, but synthesized by different methods, commonly fluoresced differently under ultraviolet light. Moreover, though the addition of certain elements seemed to activate fluorescence in sphalerite, others seemed to have the opposite effect. Explanations put forth to explain one or another of these responses were many and varied; some were later discredited and others verified as scientific understanding of the mechanisms of fluorescence in sphalerite progressed.

Natural sphalerites, too, show a wide range in color of fluorescence, with yellow, orange, and blue responses predominating *(plate 23)*. The possible combinations of potential activators in natural sphalerites, however, are so numerous that attribution of a given response to a specific impurity commonly is impossible without sophisticated and time-consuming laboratory investigation. The coveted blue-fluorescing sphalerite from Sterling Hill and Franklin is a case in point: even today the belief among some local collectors that the blue fluorescence is due to silver remains unproven. Silver was implicated because trace amounts cause blue fluorescence in synthetic sphalerite and are present in the Sterling Hill and Franklin material, but traces of numerous other impurities are present also, and the question is by no means settled. Nor has it been established that the fluorescence is due to an impurity element at all; blue fluorescence in sphalerite can also be activated by defect structures, as discussed in a later section.

Sphalerite is hardly the only example of such fluorescent complexity in the mineral kingdom; fluorapatite, fluorite, willemite, calcite, and several of the feldspars are others. Some of these are discussed in later sections of this chapter. Relative to the great variety and number of fluorescent minerals in nature, few species have been so thoroughly investigated that the activators responsible for a given response are known. Nevertheless, much has been learned in the past 60 years. In this chapter we review the three general classes of activators in fluorescent minerals—intrinsic, impurity, and defect activators—and then proceed to the dark side of this topic: the quenchers, or poisoners, of fluorescence.

Intrinsic Activators

Some mineral species nearly always show a consistent fluorescence regardless of locality. Even their pure, laboratory-prepared equivalents fluoresce in the same manner. Such minerals, for which the fluorescence is a property of the mineral itself and not of some chemical impurity or structural defect, are known as *intrinsically activated* minerals. Known and probable activators in this class of fluorescent minerals are few, but they are responsible for the fluorescence of many dozens of mineral species.

Tungstate and Molybdate Ions

By far the best known of the intrinsically activated minerals is scheelite, $CaWO_4$. The characteristic bright blue fluorescence of this species, a collector favorite, is activated by the tungstate ion, $(WO_4)^{2-}$. Similarly, the characteristic pale yellow fluorescence of powellite, $CaMoO_4$, is activated by the molybdate ion, $(MoO_4)^{2-}$. A complete chemical series extends between the two so that occurrences of molybdian scheelite and tungstenian powellite are common in nature. Most scheelite, for example, contains at least a little molybdenum, commonly a few percent or less. The shift in color of fluorescence, from blue in pure scheelite through bluish white and white to pale yellow with increasing molybdenum content, has long been used by prospectors as a quick field test for the purity of their samples. More on scheelite later—but it is worth noting here that scheelite is among the brightest and most appealing of all fluorescent minerals, and that large specimens showing attractive patterning under the lamp are eagerly sought by collectors.

Uranyl Ion

The uranyl ion, $(UO_2)^{2+}$, is the cause of fluorescence in a major group of economically important minerals, the secondary uranium species. The characteristic yellow-green fluorescence of numerous members of this group was extensively used as a prospecting guide during the uranium boom of the 1950s and early 1960s in the western United States. The widespread occurrence of these minerals (though generally in small amounts) and the economic importance attached to them made them instant classics among collectors of that period; even today the color of fluorescence of many minerals is described simply as "uranium green." In truth, however, uranium minerals exhibit a range of colors under the lamp, including shades of yellowish green through green and bluish green, and rarely even brown to orange. Again, more later.

Mercury

Paired mercury ions of unlike charge are the probable cause of fluorescence in calomel, a mercury chloride mineral best known in this country from the mercury deposits near Terlingua, Texas. The deep orange to orange-red fluorescence of this mineral under longwave ultraviolet light is quite attractive. Good specimens, always in short supply, are now uncommon on the mineral market.

Lead

The attractive yellow fluorescence of two common lead minerals, cerussite and anglesite, is self-activated. Other lead minerals showing a similar response include cotunnite, lanarkite, leadhillite, massicot, matlockite, mattheddleite, phosgenite, pyromorphite, and susannite. The brightness of fluorescence varies over a wide range among these minerals, even within the same species. Often it is of only weak to moderate intensity, but some specimens fluoresce brightly enough to display. Good specimens of cerussite and phosgenite are particularly notable in this regard.

Boron and Titanium

Other mineral species suspected of being self-activated include some of the borates and the titanium minerals. The borate minerals constitute a particularly large and geologically widespread group of minerals whose most familiar member, from a collector's point of view, is colemanite *(plate 5)*. More than 50 borate minerals have been reported to fluoresce, generally in shades of white, cream, yellow, and pale orange. Fluorescent titanium minerals, in contrast, are few—only five have been reported as fluorescent—but among them is benitoite, a barium titanium silicate, one of the most beautiful of all known fluorescent minerals. This mineral, the state gemstone of California, occurs in lustrous, well-formed blue crystals whose fine, deep blue fluorescence under shortwave ultraviolet light is rivaled by few other species for richness of color. The fluorescence is activated by titanium as Ti^{4+} ions.

Impurity Activators

Minerals activated by chemical impurities constitute by far the largest class of fluorescent minerals. Mineralogically important impurity activators include the following.

Uranyl Ion

The same uranyl ion that causes yellow-green fluorescence in numerous uranium minerals causes a similar response in several other species where it is present in only miniscule amounts as a chemical impurity. Among such species the most familiar to collectors undoubtedly is opal, particularly the nongem variety termed *common opal.* Much common opal is massive and translucent to opaque *(plates 12 and 13),* but some—the material called *hyalite*—is of almost waterlike transparency. Hyalite in many places forms films so thin that the mineral is difficult to detect by eye, yet its bright green glow is instantly recognizable at night by the light of a portable ultraviolet lamp from a distance of 10 feet or more. A similar uranyl-activated fluorescence in chalcedony, the fine-grained variety of quartz, has been reported from more than 100 localities worldwide. Adamite, a zinc arsenate mineral, is another species whose bright yellow-green fluorescence is activated by the uranyl ion; the splendid crystal groups from the Ojuela mine, Mapimí, Durango, Mexico, are collector favorites.

Manganese

Divalent manganese (Mn^{2+}) is the known or suspected cause of fluorescence in a wide variety of minerals from diverse geologic settings. Common examples include red-fluorescent calcite from countless small mines and prospects in the southwestern United States and Mexico, yellow-fluorescent fluorapatite from numerous pegmatites, and yellow- to orange-fluorescent wollastonite from chemically altered carbonate rocks near igneous plutons. Manganese in halite causes a bright orange-red fluorescence well shown by specimens from the Salton Sea in California *(plate 11)* and by the impressively large, colorless cubic crystals from the Inowroclaw mine in Poland; fine examples of either are magnificent additions to any fluorescent display. By far the best known of the manganese-activated minerals, however, are those from the zinc ores of Franklin, New Jersey. Among the ten Franklin species most coveted by collectors for their brightly colored fluorescence (calcite, willemite, esperite, wollastonite, turneaureite, barite, margarosanite, hardystonite, clinohedrite, and manganaxinite) the first five are activated by manganese, and the last two may be also.

A noteworthy trait of most of the manganese-activated minerals is their richly colored fluorescence. Many people find the bold oranges, yellows, greens, and reds of these minerals immensely appealing, which explains their prevalence in most displays of fluorescent minerals worldwide. An added benefit in many specimens is that two or more manganese-activated minerals occur together, thereby creating some fascinating combinations of color and texture under the lamp. The hobby of collecting fluorescent minerals would be impoverished immeasurably if the manganese-activated species were removed from the scene.

Chromium

Trivalent chromium (Cr^{3+}) substituting for aluminum is responsible for fluorescence in some of the most popular gem species in the United States. In ruby (the red gem variety of corundum) and in spinel, chromium in small amounts—typically 1% or less—activates a beautiful, deep red fluorescence under longwave ultraviolet light. The pink to deep red color of many specimens in daylight is likewise due to chromium; both minerals are colorless when pure. Some emeralds, too, fluoresce red owing to chromium as an impurity constituent, though in this mineral—a gem variety of beryl—the chromium imparts a deep *green* daylight color. Chromium has also been suggested as the cause of the rare red fluorescence in several other minerals: uvarovite, an emerald-green garnet prized as a gemstone; kyanite, a mineral widespread in metamorphic rocks; and topaz, another popular gem species. The chromium-activated fluorescence of all these minerals is near the long-wavelength limit of the visibility range of the human eye. Peak emission in a variety of other species occurs at slightly longer wavelengths, in the near infrared, and thus is invisible to most people.

Fluorescence in all the Cr^{3+}-activated minerals, though commonly apparent under longwave ultraviolet light, is much more effectively excited by visible light of blue or greenish-yellow color. Exposure to shortwave ultraviolet light does not result in fluorescence. The feeble glow of some rubies under conventional shortwave lamps is due instead to longwave ultraviolet and violet light, which are always produced in small amounts by these lamps.

Rare-Earth Elements

The rare-earth elements, so called because their oxides (known as *earths* to earlier generations of scientists) are indeed rare in economically significant concentrations, constitute an important class of impurity activators in minerals. In alphabetical order these elements include cerium (Ce), dysprosium (Dy), erbium (Er), europium (Eu), gadolinium (Gd), holmium (Ho), lanthanum (La), lutetium (Lu), neodymium (Nd), praseodymium (Pr), samarium (Sm), terbium (Tb), thulium (Tm), and ytterbium (Yb). The colors of fluorescence induced by these elements are highly variable because (a) several or more of them commonly are present at once, (b) some of them can occur as either divalent or trivalent ions, depending on the mineral host and its conditions of formation, and (c) interactions among coexisting rare-earth elements, and between the rare-earth elements and other impurity ions, are common. Consider, for example, the commonly fluorescent mineral fluorite, CaF_2. Experiments with synthetic fluorite have shown that any of the rare-earth elements can enter the fluorite structure, some as either divalent or trivalent ions, and that many of them can activate fluorescence in the visible wavelength range. In natural fluorite the most abundant rare-earth impurity generally is divalent europium (Eu^{2+}), which substitutes for calcium and is well known as the cause of the deep blue to violet fluorescence typical of the species. *Trivalent* europium, however, is present in some specimens and contributes to fluorescence of markedly different color: red to orange-red. In some fluorite both divalent and trivalent europium are present at once. Moreover, additional rare-earth ions can activate their own characteristic fluorescence, again altering the overall resultant color: Yb^{2+} will add a yellow-green component to the fluorescence, Dy^{3+} will add a yellow-orange component, and so on. The sum of all the visible light emitted by each rare-earth element in the fluorite determines the color we perceive as fluorescence. Given the known wide variation in rare-earth contents of natural fluorites, and the differing relative proportions of rare-earth elements from one locality to another, a wide range in color of fluorescence is only to be expected of the species. Reported colors of fluorescence for specimens from more than 350 localities include violet, blue-violet, blue, red-violet, red, pink, orange, yellow, yellow-green, blue-green, green, cream, yellow-white, bluish white, and white; most of these responses probably are due to rare-earth activators. Measurement of the wavelength distribution of the emitted light often enables researchers to identify which rare-earth activators are present, and sometimes to estimate their concentrations, by comparison to spectra of synthetic fluorites whose content of rare-earth elements is known.

The rare-earth elements are known activators of fluorescence in many other minerals besides fluorite, including some of the feldspars, various members of the apatite group, anhydrite, barite, celestite, calcite, agrellite, scheelite, and zircon. In a commercial sense the rare-earth elements are perhaps best known as activators of phosphors in color TV screens and computer monitors. In these devices, however, the bright images that you see on the screen are activated by an electron beam rather than ultraviolet light.

Disulfide Ion

The disulfide ion, S_2^-, is a known cause of fluorescence in at least eight aluminosilicate minerals from three mineral groups: *sodalite, haüyne, lazurite,* and *nosean* (sodalite group); *meionite* and *marialite* (scapolite group), and *vishnevite* and *cancrinite* (cancrinite group). The related aluminosilicate mineral *tugtupite* also owes its fluorescence to disulfide ions. The characteristic color of S_2^--activated fluorescence in these minerals ranges from yellow to orange-red, with yellow-orange to orange hues predominating *(plate 10)*. The fluorescence generally is brightest under the longwave lamp and in some specimens is very bright indeed. Meionite (commonly marketed as wernerite) from southern Quebec is one of the most intensely fluorescent minerals known, ranking third only behind some willemite and hyalite opal. Adding to their appeal, the fluorescence of several of the disulfide-activated species—notably some sodalite, meionite, and tugtupite—is of different color under the shortwave lamp. Tugtupite shows the most extreme color change, from salmon orange under longwave ultraviolet light to deep red *(plate 6)* under shortwave ultraviolet light.

Disulfide impurity ions are responsible for another remarkable property of some of the aluminosilicate minerals: *tene-*

brescence, or reversible photosensitivity. Some specimens of the hackmanite variety of sodalite display it exceptionally well: the sodalite, normally white to pale pink in daylight, deepens in color to dark purplish red upon exposure to ultraviolet light. The color then gradually bleaches in daylight or incandescent light and returns to normal in a matter of minutes or hours. The cycle of darkening and bleaching can be repeated indefinitely.

Defect Activators

Crystal structures of minerals are depicted in many textbooks in idealized form, as perfect three-dimensional arrays of interlinked ions, each part of the crystal exactly like every other part. Figure 1 shows such an idealized structure for the mineral *halite,* NaCl. The idealized form is appropriate for showing the types of ions present, the nature of the bonds between them, and the manner in which they are arranged relative to one another in space, but one should keep in mind that no real crystal conforms exactly to this ideal. Real crystals, whether natural or manmade, contain a variety of defects. Some crystals, whose structures should contain orderly rows of ions like soldiers on parade, in places contain partial rows wedged between the complete ones, resulting in local distortion *(fig. 2).* Others, built up from sheets of ions stacked one

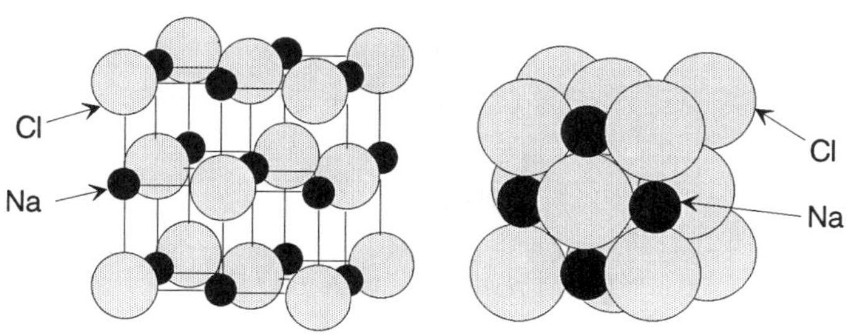

Figure 1. The crystal structure of halite, NaCl. Left, "exploded" view showing Na-Cl bonds; right, close-packed model conforming to actual structure.

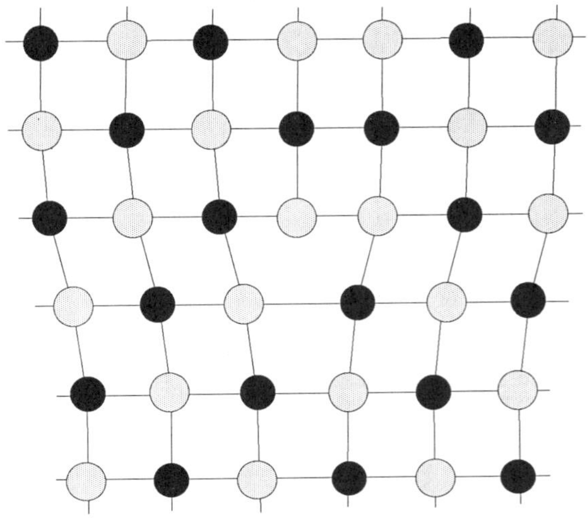

Figure 2. An example of a line defect in a crystal structure.

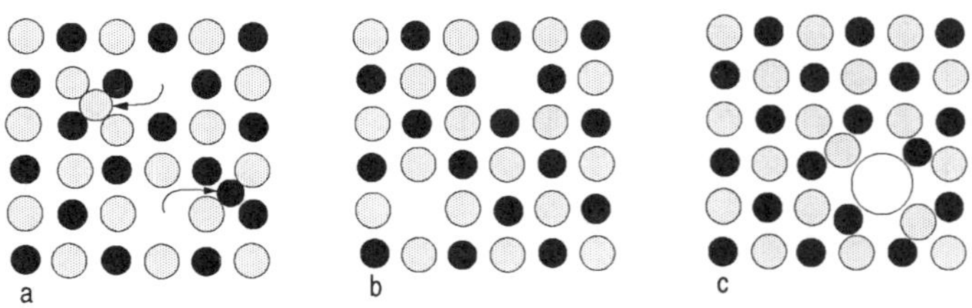

Figure 3. Three types of point defects in crystals: A, Frenkel (mislocation) defect; B, Schottky (vacancy) defect; C, interstitial (impurity) defect.

upon another, contain stacking faults that result in local mismatches and distortion of the ideal structure. The defects that most influence mineral fluorescence, however, are those known collectively as *point defects.* Among these are *Frenkel defects,* which are ions in abnormal locations instead of their proper sites in the crystal structure *(fig. 3a); Schottky defects,* which are vacancies due to missing ions *(fig. 3b);* and *interstitial defects,* which are impurity ions in crystal sites not normally occupied *(fig. 3c).* The electron cloud around any ion located near one of these defects is distorted, and thus the crystal absorbs and dissipates light energy differently within and near a defect than within structurally "perfect" parts of the crystal. Point defects are a common cause of color in minerals that in daylight would ordinarily be colorless or white; they are also a common cause of fluorescence. Defect structures are associated with yellowish-green fluorescence in diamond, with blue fluorescence in sphalerite, and with many examples of white or near-white fluorescence in calcite. The fluorescence of some fluorite and of various sulfate minerals is also activated by crystal defects.

An important aspect of Frenkel defects *(fig. 3a)* is that their presence requires no deviation from the ideal composition of the mineral; the defect is simply a normal ion in an abnormal position. Schottky defects, too, may leave mineral compositions unchanged because they commonly occur in pairs. If, for example, a zinc vacancy in sphalerite were located near a corresponding sulfur vacancy, not only would local electrical neutrality of the crystal be preserved, but also its bulk composition would remain unchanged. Thus, even the most detailed chemical analyses may furnish no clue as to the cause of fluorescence in minerals where structural defects are involved. Worse, impurity elements that have nothing to do with activating the fluorescence, or that play only an indirect role as explained in a later section on promoter ions, may mistakenly be suspected as the activator until detailed structural and experimental studies reveal otherwise. The study of defect-activated fluorescence is a challenging one, and much remains to be learned.

Quenchers of Fluorescence

Certain elements, notably iron, nickel, copper, and cobalt, are effective quenchers, or poisoners, of fluorescence. Among 566 mineral species reported as fluorescent in a 1989 compendium, only four contain nickel or cobalt in their chemical formulas. Nickel is especially effective in suppressing fluorescence in many species; in amounts of only 1-10 parts per million in sphalerite it results in a marked decrease in brightness, and at slightly higher concentrations it quenches the fluorescence of willemite. Iron has a similar effect. The quenching effect of copper, too, is obvious in many minerals, as in the list below for members of the autunite group.

Species	**Chemical formula**	**Fluorescence**
Autunite	$Ca(UO_2)_2(PO_4)_2 \cdot 10\text{-}12H_2O$	Bright yellow-green
Saléeite	$Mg(UO_2)_2(PO_4)_2 \cdot 10H_2O$	Bright yellow-green
Torbernite	$Cu(UO_2)_2(PO_4)_2 \cdot 8\text{-}12H_2O$	*Nonfluorescent*
Heinrichite	$Ba(UO_2)_2(AsO_4)_2 \cdot 10\text{-}12H_2O$	Bright yellow-green
Nováčekite	$Mg(UO_2)_2(AsO_4)_2 \cdot 12H_2$	Moderate yellow-green
Zeunerite	$Cu(UO_2)_2(AsO_4)_2 \cdot 10\text{-}16H_2O$	*Nonfluorescent to weak yellow-green*

Yet copper as a trace impurity in some minerals, notably sphalerite, can *cause* fluorescence, and the mineral marshite (copper iodide), despite containing fully one-third copper by weight, nevertheless fluoresces bright red under the longwave lamp—and so once again we are reminded of the remarkable complexities of fluorescence in minerals.

Some elements can act as quenchers of fluorescence in some minerals but as activators in others, depending in part on oxidation state. Iron is the best known and most common of these. Nearly all minerals containing *divalent* iron (Fe^{2+}) as an essential constituent do not fluoresce. Even as an impurity constituent, in amounts of only 1% or less, divalent iron has a similar effect; the feebleness of the fluorescence of many mineral species is due not to a deficiency of the activator but to an excess of iron. *Trivalent* iron (Fe^{3+}), in contrast, quenches fluorescence in many minerals but activates it in some; in orthoclase feldspar and perhaps also scapolite it results in deep red fluorescence.

The Role of Coactivators in Impurity-Activated Minerals

Two conditions must be fulfilled for any mineral to fluoresce: the mineral must absorb some of the ultraviolet radiation beamed upon it, and the absorbed energy, or part of it, must be available to the activator instead of being dissipated within other components of the crystal. Minerals transparent to ultraviolet radiation, much as window glass is transparent to visible light, cannot fluoresce.[3] The presence of a potent activator such as divalent manganese may not be sufficient to induce fluorescence; by itself, the Mn^{2+} ion is a most ineffective absorber of ultraviolet energy. For this reason, many specimens of calcite, even those containing appreciable manganese, fluoresce only feebly. Bright fluorescence in such minerals depends on the presence of a second impurity, one that will absorb ultraviolet radiation, transfer some of the absorbed energy to the activator ions, and hence allow the activator to perform. Such additional impurities are called *coactivators,* or *sensitizers.*

The discovery of coactivators in some minerals came about almost by accident. In the mid-1940s, two researchers with the U.S. Geological Survey embarked upon a series of simple experiments whose results at first seemed confusing but that later provided the key to understanding red fluorescence in halite. Halite, NaCl, is a fairly common mineral best known to most people as table salt. The two scientists examined red-fluorescing halite from several localities and found that the spectrum of the emitted light was similar in all of them, suggesting a uniform cause of fluorescence. Moreover, the halite spectra much resembled those of red-fluorescing calcite from Franklin,

---

[3] One cannot tell if a mineral transmits or absorbs ultraviolet light merely by looking at it. Some minerals that are transparent in daylight, such as cerussite, anglesite, and uranyl-activated hyalite opal, nevertheless are strong absorbers of ultraviolet light.

New Jersey, which was known to be activated by manganese. Chemical analyses disclosed the presence of manganese in all of the natural halite samples, thereby settling the question—or did it?

Fascinating complications arose as the scientists dissolved the natural crystals in water and then slowly evaporated the salt solution, allowing the halite crystals to reform. Time and again, the only crystals that showed bright red fluorescence were those that grew during the initial stages of evaporation. Halite that formed later, either as new crystals or as enlargements of previous ones, showed only a feeble response or none at all, even though the solution from which they were precipitating still contained abundant manganese. Here, then, was a critical clue: manganese alone was not sufficient to cause fluorescence. The two workers then hypothesized that a *second* substance, one readily incorporated in early crystals and thus no longer available for later ones, must act in concert with manganese to cause the red fluorescence of halite. Additional chemical analyses provided the answer: it was lead, present in minute amounts as an impurity in the early formed, fluorescent crystals but missing from those formed later. As an additional check on the results, a few drops of lead nitrate solution were added to a manganese-bearing brine from which nonfluorescent halite crystals had been forming; the lead-enriched solution then began depositing bright red-fluorescent crystals! The role of lead as a coactivator of red fluorescence in halite now stood well established.

The potency of lead as a coactivator was graphically illustrated by accident during these investigations. During one experiment, a small amount of lead-depleted salt solution spilled over the side of a beaker and dried. Later, when an ultraviolet lamp was beamed on this beaker, the word *Pyrex* (a brand name) glowed bright red as it never had before. It was later learned that the brand name had been etched into the glass with a flux of lead borate. The salt solution had absorbed sufficient residual lead from the glass that the tiny crystals that grew on it—those that coated the word *Pyrex* on the beaker—were activated into bright red fluorescence.

As with activators, the list of known and potential coactivators in minerals is a long one, including aluminum, antimony, arsenic, bismuth, copper, erbium, gallium, indium, lead, manganese, thallium, tin, and various rare-earth elements. Lead is one of the most potent of these, both because it is a widely distributed element in the Earth's crust and because only small amounts are needed to cause some minerals to absorb ultraviolet light strongly.

The Role of Promoter Ions in Defect-Activated Minerals

Experiments with synthetic sphalerite have shown that blue fluorescence can be activated not only by silver, as noted before, but also by zinc vacancies. The formation of vacancies in the crystal structure is promoted by any of several impurity elements, including Cl^-, Br^-, or I^- substituting for S^{2-}; or alternatively by Ga^{3+} or Al^{3+} substituting for Zn^{2+}. Each of these substitutions results in an excess of positive charge, which is compensated by omission of some of the zinc ions in the structure. The actual emission of light that we see as fluorescence is associated with the zinc vacancies and thus is a true defect-activated fluorescence. The impurity ions play no direct role in the process of fluorescence; they merely stimulate the formation of the zinc vacancies. For this reason they are sometimes termed *promoter ions,* a term that serves to distinguish their role from the very different function performed by coactivators in impurity-activated fluorescent minerals.

There are many examples in the mineralogic literature of substitutional impurities resulting in a charge imbalance in the mineral structure and thereby inducing the formation of ion vacancies. In some, as in the example above, the impurities serve only as promoter ions. In others, however, the impurity ion and an adjacent vacancy participate together in activating fluorescence. Such complex centers of fluorescence blur the line between impurity and defect activators.

The Effect of Crystal Structure on Color of Fluorescence

One might suppose, or at least hope for the sake of simplicity, that a given impurity ion would activate a fairly consistent color of fluorescence in all minerals. In practice this is not always so, but among those that do are some of the most important impurity activators in minerals: the trivalent rare-earth ions. The electrons responsible for fluorescence in these ions are shielded by other electrons and are not involved in chemical bonds with surrounding ions; hence their energy levels (their "orbits", if you like) remain relatively little affected by variations in crystal structure. The response of these shielded electrons to ultraviolet radiation is much the same regardless of the nature of, and distances to, nearby ions in the mineral host. Similarly, many secondary uranium minerals, as already discussed, fluoresce some shade of green, and the color differences among most of them become apparent only if they are examined side-by-side. The colors of fluorescence of uranyl-activated hyalite opal, chalcedony, and adamite are closely similar. It matters little in these minerals whether the uranyl ion is present as a trace impurity or major constituent, or whether the mineral is a phosphate, arsenate, silicate, sulfate, or oxide; the color of fluorescence in most of them varies only within narrow limits.

In some other activator ions the electrons responsible for fluorescence are the same as those involved in chemical bonds with surrounding ions in the crystal structure. Such activators are highly sensitive to the geometry of the crystal site in which they occur, and thus they activate fluorescence of different color in different minerals. By far the most effective and versatile of these is divalent manganese.

The wide range in colors of fluorescence among the manganese-activated minerals—from red through orange and yellow to green, or more than ⅔ of the visible spectrum—is a direct reflection of the diverse crystal structures of the numerous fluorescent minerals in which manganese is present as an impurity element in substitution for calcium or zinc. In calcite, for example, each manganese ion substituting for calcium is surrounded by six oxygen ions to form, in effect, an octahedral cage with manganese in the center *(fig. 4a)*. The calcite fluoresces orange-red. A similar octahedral arrangement exists in halite (NaCl), wherein each manganese impurity ion is surrounded by six chloride ions; here too the mineral fluoresces orange-red. In willemite, however, each manganese impurity ion is surrounded by *four* rather than six oxygen ions, and thus the resultant site has a tetrahedral rather than an octahedral configuration *(fig. 4b)*. The large difference in structural environment is reflected in an equally large shift in the color of fluorescence, now yellow-green rather than orange-red. A similar effect is seen in esperite, whose lemon-yellow fluorescence, only slightly different in color from the yellow-green of willemite, is likewise caused by divalent manganese substituting for zinc in a tetrahedral site. Manganese is also the known or suspected activator in clinohedrite (fluorescent orange), wollastonite (yellow to orange), manganaxinite (red), benstonite (red), some sphalerite (orange) and fluorite (green), and various members of the amphibole, pyroxene, and apatite groups. In all these minerals the immediate environment of the manganese ion exerts a profound effect on the color of fluorescence.

Divalent europium offers a more subtle example of variable fluorescence in different minerals. Its fluorescence in

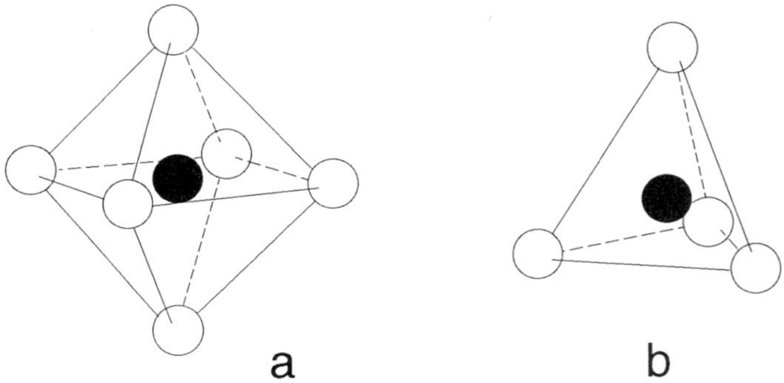

Figure 4. A manganese ion in a 6-coordinated (octahedral) site (left) and a 4-coordinated (tetrahedral) site (right).

oligoclase peaks at 390 nm, just below the shortest wavelengths of light visible to most humans, but one side of the peak extends slightly into the visible range; thus a dull violet fluorescence is characteristic. In fluorite the europium emission peak is shifted to the violet and blue region, centered at about 420 nm, and in fluorapatite it is shifted still further, into the blue and greenish blue at about 460 nm. These shifts, though not large, are readily noticeable if examples of all three species are placed side-by-side.

Just as different colors of fluorescence can be caused by the same activator, so too can seemingly identical colors of fluorescence be caused by *different* activators. To show this we turn to a spectrofluorometer, a machine that measures the brightness of fluorescent light as a function of wavelength and plots the results. The resultant plot for any given specimen is called its *emission spectrum.* Two such plots, one for willemite and one for common opal, are shown in figure 5. The willemite is activated by manganese and the opal by uranyl ion, yet their colors of fluorescence appear to our eyes as virtually identical. Their emission spectra, however, are dramatically different—a smooth, bell-shaped curve for the willemite versus a more complex curve with multiple peaks for the opal. Despite the

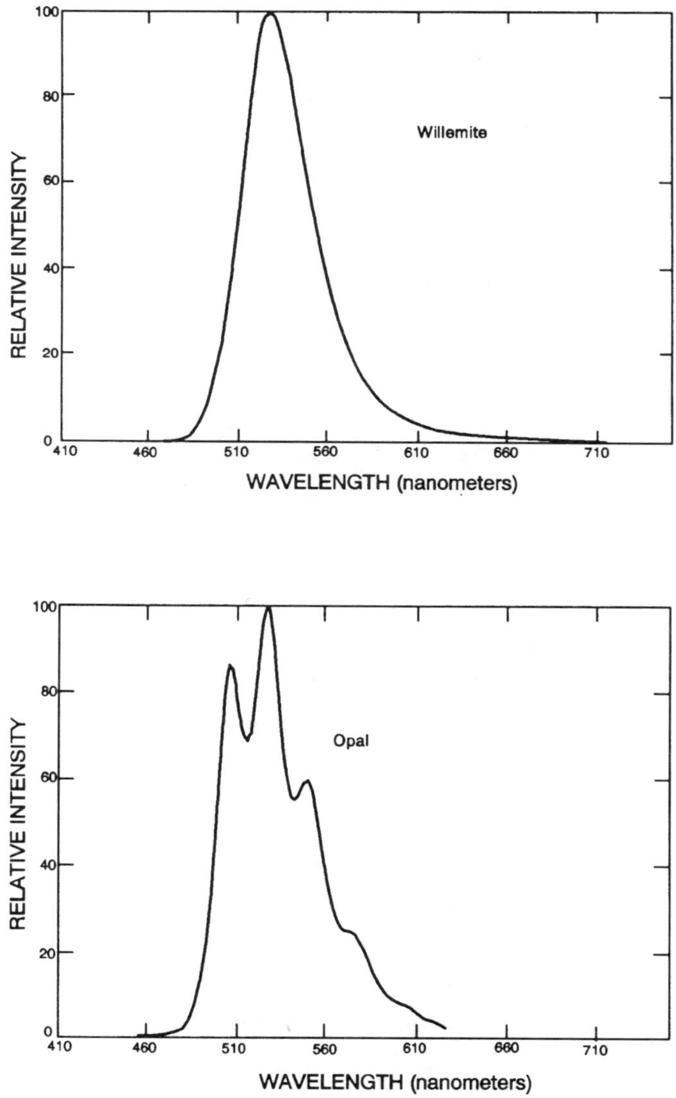

Figure 5. Fluorescence emission spectra of willemite (top) and common opal (bottom) (from Newsome and Modreski, 1981, Journal of the Fluorescent Mineral Society, vol. 10, no. 1, p. 7-56). The fluorescence of these two minerals as perceived by the human eye is of nearly identical yellowish-green color, but the different shapes of the emission spectra show that different activators are involved.

variable colors of fluorescence caused by some activators, as discussed above, the shapes of their emission spectra commonly are characteristic. Emission spectra often enable researchers to narrow the list of suspected activators considerably and so concentrate on one or two remaining possibilities.

Interactions in Complexly Activated Minerals

All minerals contain structural defects, and few are chemically pure. Though the effects of various impurities and defects may merely be additive, as in two activators that each produce their own separate fluorescence in a mineral, in many cases they are interactive: the presence of one influences the fluorescence properties of the other. Only a few examples are given here to illustrate some of the possibilities.

(1) The divalent rare-earth elements (RE^{2+}) are strong absorbers of ultraviolet light and effective activators of fluorescence. The trivalent rare-earth elements (RE^{3+}), however, absorb ultraviolet light only weakly and require the presence of a suitable coactivator to fluoresce well. When both are present simultaneously it is common for energy transfer to take place from RE^{2+} ions to RE^{3+} ions, thereby stimulating strong fluorescence in the latter. The RE^{2+} ions in this role are coactivators.

(2) The yellow fluorescence of Mn^{2+} in fluorapatite is suppressed or eliminated in the presence of trivalent samarium (Sm^{3+}) due to energy transfer between them. The pink fluorescence of the Sm^{3+} is correspondingly increased. Activation of one type of ion thus takes place at the expense of another, whose fluorescence is quenched.

(3) Though quenching of fluorescence generally results from interaction between unlike ions, self-quenching (the phenomenon known as *concentration quenching)* is also known.

In chromium-activated ruby corundum, the light emitted by some of the Cr^{3+} ions is partially reabsorbed by other Cr^{3+} ions, thereby limiting the overall brightness of fluorescence. Through a different but equally effective process, the fluorescence of minerals activated by some of the rare-earth ions decreases and ultimately becomes extinguished with increasing concentrations of the activator. Thresholds for concentration quenching are different for different activators: 1% or less for most Cr^{3+}-activated minerals, but several tens of percent for some of the minerals activated by rare-earth ions. Concentration quenching is the reason that many mineral species do not fluoresce even though they contain large amounts of a well-known activator.

(4) As noted before, strong fluorescence in ions that are weak absorbers of ultraviolet light most commonly is stimulated by interaction with a suitable impurity coactivator, as in the fluorescence of Mn^{2+} coactivated with Pb^{2+} in calcite. In some minerals, however, the crystal itself can function as the coactivator, and no additional impurity element is needed. Willemite is one example, wherein strong yellow-green fluorescence is made possible by interaction between Mn^{2+} ions and the crystal lattice.

Summary

The material presented in preceding pages only touches upon a fascinating subject that can become a lifelong hobby. Fluorescence in minerals can occur in many ways. For some species the process is well known, but for many others it is incompletely understood, and much remains to be discovered. A few key points, however, are well documented:

(1) Fluorescence is an intrinsic property of some minerals, but in many others it results from either chemical impurities or structural defects.

(2) Some impurity elements that activate fluorescence can-

not effectively function alone but require the presence of a coactivator. The coactivator most often is another impurity element but in some cases is the same element in a different oxidation state, or is the crystal lattice itself. The coactivator participates directly in the process by transferring energy to the activator, thereby causing it to fluoresce.

(3) Fluorescence in some minerals activated by crystal defects depends on the presence of promoter ions in suitable concentration. Unlike coactivators, the promoter ions do not participate directly in the process of fluorescence but merely stimulate the formation of the defects.

(4) A few impurity elements—notably copper, cobalt, iron, and nickel—are effective quenchers of fluorescence in almost all minerals in which they occur.

(5) The role played by a given impurity ion—whether as activator, coactivator, promoter, or quencher—commonly is not fixed but depends on the nature of the mineral host, the identity of other impurities present, and their concentrations. An element that would normally be an activator may instead become a coactivator in the presence of additional impurities. Two activators present simultaneously may interact so that one suppresses or quenches fluorescence in the other. A given impurity element may activate fluorescence in one mineral but not another, and may quench fluorescence in a third. Other possibilities are numerous.

Activators in Fluorescent Minerals

Earl R. Verbeek

Part 2: Some Notable Fluorescent Minerals and Mineral Groups

Introduction

Some of the major mineral groups important to the collector of fluorescent minerals are discussed below. The list is hardly an all-inclusive one and is meant only to introduce the reader to the richness of the fluorescent-mineral hobby and to some of the species that might enjoyably be collected. In reading these passages, please keep in mind that more than 500 mineral species from more than 6,500 localities are known to fluoresce, and that no single book can do justice to them all. Readers wishing more information than that presented here are encouraged to consult some of the sources listed in the bibliography at the end of this chapter.

Scheelite-Powellite Series

Reference to the scheelite-powellite series has been made repeatedly in this book, and for good reason. Scheelite, prized by collectors for its bright blue to white or pale yellow fluorescence, is the most important ore mineral of tungsten in the United States. Most commercial deposits of scheelite are in contact-metasomatic skarns—masses of silicate rock (garnet, pyroxene, amphibole, vesuvianite, etc.) that have replaced carbonate rock (limestone or dolostone), generally in the vicinity of a mass of granite or similar igneous rock. The heat and the additional chemical components necessary for the transformation were supplied by the granite, which intruded the original carbonate and released chemically reactive fluids as it cooled from a viscous liquid to the solid rock we see today. A few scheelite deposits are hydrothermal (deposited from hot aqueous solutions), and still others are residual placer deposits (stream sediments enriched in heavy ore minerals as particles of lighter minerals were washed away). The bright and dependable fluorescence of this mineral is indeed fortunate, for scheelite in many of its manifestations is most inconspicuous. Commonly it occurs in small (½ inch or less), irregular, white to gray or pale yellow grains resembling common quartz. In this form it is easily overlooked unless there is so much of it scattered throughout the rock that its weight becomes conspicuous: scheelite is heavy, more than twice as heavy as the other minerals with which it is most commonly associated. With a shortwave ultraviolet lamp, however, even tiny amounts of scheelite can be detected instantly.

Collectors in the western United States, particularly in California, Nevada, and Arizona, have grown accustomed to finding scheelite during their nighttime forays with the ultraviolet lamp. In small amounts it is a widespread mineral, and specimens showing minute, scattered grains are known from hundreds of localities. Large masses, however, are distinctly less common. Specimens combining large size, bright fluorescence, and intriguing texture under the lamp *(plate 7)* are among the most appealing additions to any collection of fluorescent minerals. Scheelite in good crystals is even less common, though conversely these are more likely to be available at mineral shows than is the massive material; the mineral community at large is fixated on crystals.

The fluorescence of scheelite, though nearly always evident, ranges in intensity from bright to fairly dull, the latter probably due to impurities that partially quench the fluorescence. The color of fluorescence, too, is variable. As noted previously, it varies with molybdenum content: at zero weight percent Mo (pure scheelite) the color is medium blue; at only 0.1 weight percent Mo the color has noticeably lightened to pale blue; at about 0.8 percent it is nearly white; and at about 3 percent it is pale yellow. The yellow color of fluorescence deepens slightly as more molybdenum enters the structure, but beyond about 5-6 weight percent Mo it shows little change. The large change in color of fluorescence with small amounts of molybdenum was long the basis for judging the approximate purity of scheelite ores in the field through use of a scheelite analyzer card *(fig. 6)*. Purity was important during the heyday of scheelite mining in the United States, for too much molybdenum would exact a penalty at the mill and could spell the difference between profit and loss for the miner.

Scheelite from numerous localities is compositionally zoned. Single crystals showing multiple growth layers, some fluorescing blue and others white to pale yellow, reflect varying proportions of molybdenum incorporated into the crystal during growth. Rare-earth activators in scheelite sometimes add a pink component to the fluorescence, which when superimposed on the normal color of fluorescence results in pinkish-blue to pink or pinkish-white hues. These too commonly are evident only in selected zones of individual scheelite grains or crystals.

Powellite, $CaMoO_4$, is much less common than scheelite and less common still as attractive specimens. In many localities it occurs as an alteration product of molybdenite and is found in small amounts as dusty coatings and tiny disseminated grains. Display-worthy powellite in coarse grains has

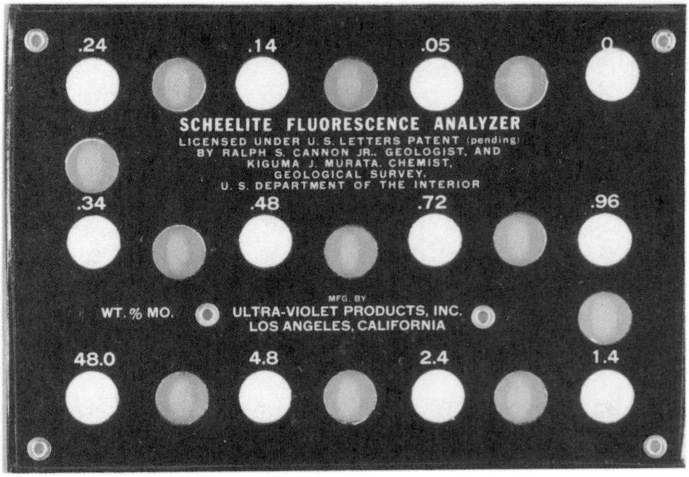

Figure 6. The scheelite fluorescence analyzer card, once marketed by Ultra-Violet Products, Inc. (San Gabriel, California) under license from the U.S. Geological Survey. Twelve samples of powdered synthetic scheelite containing known amounts of molybdenum show the gradual change in fluorescence color from medium blue to pale yellow with increasing molybdenum content. Next to each powdered sample is a hole through the card so that ultraviolet light can be beamed on a specimen of natural scheelite held beneath. The card enabled prospectors to judge the purity of scheelite ores instantly in the field.

been found at few localities and hardly ever in abundance, with rare exceptions. The most remarkable occurrences are in the basaltic lava flows of India. From several quarries near the city of Nasik have come spectacular powellite *crystals,* transparent, sharply formed, straw yellow to medium brown in color, some more than one inch long, and brightly fluorescent. Their association with lustrous, blocky, colorless to white crystals of apophyllite on black matrix adds considerably to their visual appeal.

One final note. On p. 174 of Sterling Gleason's original (1960) book he stated emphatically that scheelite does not fluoresce under the longwave lamp and is never visibly phosphorescent. With modern lamps, however, we can now see that some scheelite does indeed fluoresce under longwave ultraviolet light, commonly faint brown or orange but in some specimens moderate pale yellow. The brightness of this response increases with increasing molybdenum content; molybdian scheelite and powellite from some localities fluoresce nearly as brightly under the longwave lamp as under the shortwave lamp. Many specimens in the scheelite-powellite series show persistent phosphorescence as well, particularly after exposure to shortwave ultraviolet light, though long exposure (one minute or more) to the lamp and dark-adapted eyes are needed to fully appreciate this effect.

Apatite Group

The apatite-group minerals are popular among enthusiasts of mineral fluorescence and occur at hundreds of localities in diverse types of deposits worldwide. Members of this group are common accessory minerals in many types of igneous rocks, in the pegmatites derived from them, and in metamorphic rocks, particularly impure limestones recrystallized to marble. Most apatite species fluoresce some shade of yellow, and at least four species occur in masses of sufficient size and fluorescent brightness that they are widely sought by collectors of fluorescent minerals. Those members of the apatite group reported as fluorescent include:

Carbonate-fluorapatite	$Ca_5(PO_4,CO_3)_3F$	Phosphate apatites
Carbonate-hydroxylapatite	$Ca_5(PO_4,CO_3)_3(OH)$	
Chlorapatite	$Ca_5(PO_4)_3Cl$	
Fluorapatite	$Ca_5(PO_4)_3F$	
Hydroxylapatite	$Ca_5(PO_4)_3(OH)$	
Pyromorphite	$Pb_5(PO_4)_3Cl$	
Hedyphane	$Pb_3Ca_2(AsO_4)_3Cl$	Arsenate apatites
Johnbaumite	$Ca_5(AsO_4)_3(OH)$	
Mimetite	$Pb_5(AsO_4)_3Cl$	
Svabite	$Ca_5(AsO_4)_3F$	
Turneaureite	$Ca_5[(As,P)O_4]_3Cl$	
Vanadinite	$Pb_5(VO_4)_3Cl$	— Vanadate apatite

Among these species, the most significant to collectors of fluorescent minerals are fluorapatite, a common mineral; and svabite, turneaureite, and johnbaumite, all rare.

Fluorapatite

Fluorescence in some of the apatite-group minerals, fluorapatite in particular, has been the subject of much study because their synthetic counterparts are commercially important phosphors. The stark glow from a cool-white fluorescent light tube, for example, comes from a coating of synthetic fluorapatite activated by manganese and antimony. The yellow fluorescence of much natural fluorapatite (and probably of most other yellow-fluorescing apatites as well) is activated by manganese, as shown by distinctive emission spectra combined with chemical analyses, and also by preparation of the analogous phosphors in the laboratory. Other colors of fluorescence commonly are due to the rare-earth elements; blue and violet fluorescence has been attributed to cerium (Ce^{3+}) and europium (Eu^{2+}), and various hues of pink, violet-pink, and yellowish pink to samarium (Sm^{3+}) and dysprosium (Dy^{3+}). Fluorapatite is a widely distributed mineral, and the potential for acquiring attractive specimens is high. As of this writing, for example, good specimens of massive fluorapatite can still be collected (with permission) from the dumps of the Harding mine near Dixon, Taos County, New Mexico, and from several pegmatites in New England. Also, beautifully fluorescent crystallized specimens of fluorapatite occasionally are available from dealers at major mineral shows. Glassy crystals from the tungsten mines of Panasqueira, Portugal and from the pegmatite gem mines of the Gilgit region in northern Pakistan are particularly prized, the former for their soft blue and pink hues in some specimens, and the latter for their bright golden-yellow fluorescence.

Svabite and Turneaureite

Svabite, the namesake of Swedish mining official Anton Svab (1703-1768), is known from several iron and manganese mines in northern Sweden. From the Långban mines have come outstanding specimens showing rich layers and streaks of svabite in calcite, fluorescent bright yellowish orange and bright orange-red, respectively, but these are uncommon on the U.S. mineral market today. More familiar to most collectors is svabite from Franklin, New Jersey, but here there is a catch—few, if any, of the specimens so labeled are truly svabite. Much of it, the bright orange-fluorescing, pale greenish-gray apatite commonly pictured in older texts, is instead turneaureite, the chlorine analogue of svabite. The rest of it, particularly material showing blue to green coloration in daylight and fluorescing more weakly than turneaureite, is mostly arsenian fluorapatite. Its identification as svabite in past years was based on the assumption that arsenate apatites should be found in the orebody (where other arsenic-bearing species are common) but that phosphate apatites should be found in the surrounding marble. That assumption, tenuous to begin with, has since been disproved.

Turneaureite at Franklin is known in masses as much as four inches across and most commonly is associated with salmon-colored calcite and brown garnet. The shape and width of the emission peak of turneaureite suggest strongly that manganese is the activator. Good specimens under the shortwave lamp are spectacular: the peach-colored fluorescence of the turneaureite balances perfectly the orange-red of the calcite, forming a pleasing and unusual combination *(plate 19)*.

Johnbaumite

Johnbaumite probably is the rarest of the brightly fluorescing apatites from Franklin and Sterling Hill. One of the best specimens known consists of a large (1.5 inch), orange-fluorescent mass enclosed in red-fluorescent calcite with a comparably sized mass of cream-fluorescent barite nearby. This and similar specimens came from a deep and now inaccessible level at Sterling Hill; in recent years another excellent specimen was found on an old dump of Franklin mine rock. Few fine specimens of *confirmed* johnbaumite are known, but there remains the potential for more as apatites from other finds are analyzed.

Other Apatites

Other members of the apatite group generally fluoresce more dimly and infrequently. In daylight, however, three of them—pyromorphite, vanadinite, and mimetite—are among the most colorful and beautifully crystallized of all known mineral species and have a distinct beauty all their own. The best specimens of pyromorphite, such as those from the Bunker Hill mine in Idaho and the Les Farges mine in France, rank among the finest nongem mineral specimens in existence.

Calcite, Dolomite, and Aragonite Groups

Minerals of the calcite, dolomite, and aragonite groups are all carbonates of relatively simple chemical composition, but the groups differ from one another in some aspects of their crystal structure. Selected members of all three groups are standard items in thousands of mineral collections, whether of fluorescent minerals or not. In the following list, numbers in parentheses preceding each mineral name are the number of localities from which the mineral has been reported to fluoresce, as listed in the 1989 edition of the *Journal of the Fluorescent Mineral Society* and modified by more recent data.

In part, the wide range in numbers reflects the effectiveness of several poisoners of fluorescence. Among members of the calcite group, for example, it is no accident that the carbonates of calcium (calcite), magnesium (magnesite), and zinc (smith-

Calcite Group

(962)	Calcite	$CaCO_3$
(0)	Gaspéite	$NiCO_3$
(17)	Magnesite	$MgCO_3$
(2)	Otavite	$CdCO_3$
(9)	Rhodochrosite	$MnCO_3$
(1)	Siderite	$FeCO_3$
(33)	Smithsonite	$ZnCO_3$
(1)	Sphaerocobaltite	$CoCO_3$

Dolomite Group

(3)	Ankerite	$Ca(Fe,Mg,Mn)(CO_3)_2$
(36)	Dolomite	$CaMg(CO_3)_2$
(1)	Kutnohorite	$Ca(Mn,Mg,Fe)(CO_3)_2$
(0)	Minrecordite	$CaZn(CO_3)_2$
(1)	Norsethite	$BaMg(CO_3)_2$

Aragonite Group

(195)	Aragonite	$CaCO_3$
(38)	Cerussite	$PbCO_3$
(38)	Strontianite	$SrCO_3$
(14)	Witherite	$BaCO_3$

sonite) are commonly reported as fluorescent, whereas the carbonate of iron (siderite) is not. The near absence of fluorescence in siderite is hardly attributable to rarity of the species; siderite is a widespread and common mineral. Similarly, ankerite, gaspéite, and sphaerocobaltite all contain well-known quenchers of fluorescence and are rare or unknown as fluorescent minerals, though for the last two, genuine rarity of the species is a large contributing factor. Probably most reports of fluorescence in these species are in error. Likely causes include misidentification (common among carbonates, where crystal forms are similar and color is an unreliable guide), and mistaking the fluorescence of a thin coating or an included mineral for that of the dominant species.

Calcite and aragonite, both reported as fluorescent from hundreds of localities, clearly stand apart from the rest of the carbonates and deserve extended mention here. Both species commonly are represented in museum displays of fluorescent minerals, show highly variable fluorescence, and are worthy of close study.

Calcite

The occurrence of calcite in large, sharply formed crystals, its diversity of colors and mineral associations, its common jewellike transparency, and its fluorescence are only some of the reasons this species is so coveted by mineral collectors, some of whom collect nothing else. Add to this its commonly modest price, and it is easy to see why calcite is well represented in thousands of mineral collections worldwide. The status of calcite as a major fluorescent species is unquestioned, and the color range of its fluorescence is broader than that of any other mineral except fluorite. Some of the more important fluorescent varieties are discussed below.

Red-Fluorescent Calcite

Calcite activated by manganese and showing orange-red to red fluorescence is familiar to many people who have combed old mine dumps at night with a portable ultraviolet lamp. Specimens on the mineral market commonly are labeled "manganocalcite," an informal (and incorrect) term for manganoan calcite. Though often described simply as "red," the fluorescence of manganese-activated calcites is actually quite variable. Specimens assembled from diverse localities and examined as a group almost always show a dominantly orange-red fluorescence, but commonly among them are some that fluoresce a true red color, or pink, or purple, so that an entire range of reds is represented. Some localities, such as Franklin and Sterling Hill, have produced calcite of highly variable fluorescent color from the same orebody. Variability is shown in other ways as well: much manganese-activated calcite responds best to the shortwave ultraviolet lamp but some does not, and the brightness of fluorescence (even at single localities) can range from vivid to nearly imperceptible.

Though synthetic manganese-activated calcites have been well studied in the laboratory, their more complex natural counterparts are relatively poorly understood. Much of the variability in fluorescence of natural calcites probably is due to the presence of different coactivators and to concentration effects. Additional variability, especially in color of fluorescence, commonly is due to the superimposed fluorescence caused by other activators in addition to manganese. In some specimens one can detect three different fluorescent responses, all active at once. Regardless of color under the lamp, however, a brief orange-red phosphorescence, so fleeting that close attention is required to detect it, is characteristic of manganese-activated calcites. This momentary phosphorescence, mistakenly referred to as a "flash" in some of the hobbyist literature, varies considerably in initial brightness from one specimen to another.

Though massive, bright red-fluorescing calcite is attractive in its own right, specimens combined with a second or third fluorescent mineral offer more interesting possibilities. The most varied of these, as noted often before in this book and illustrated on some of its pages, come from the remarkable zinc deposits of Franklin and Sterling Hill. Calcite from these deposits forms a stunning backdrop to numerous other fluorescent species, most of them of sharply contrasting colors under the lamp. Unassuming in daylight—much of the calcite is massive and pale gray to white—the mineral comes to life under shortwave ultraviolet light in colors ranging from brilliant orange-red to violet, sometimes mixed in a single, texturally complex mass. Specimens combined with green-fluorescing willemite are the most common and show an almost infinite variety of form. Similar combinations of calcite with willemite are known from a surprising number of other localities, though not much of it approaches the quality of the best Franklin-Sterling Hill material. Elsewhere in the world red-fluorescing calcite is found in attractive combinations with other minerals: coatings of bright blue-fluorescing hydrozincite; masses of fluorite, with their deep blue-violet glow; sphalerite grains which fluoresce orange; scheelite, with its intense blue to white and even yellow fluorescence; or hyalite opal, with its bright yellow-green fluorescent coatings. And from the Långban mines in northern Sweden, some exquisite rarities: barytocalcite (fluorescent pale blue), svabite (bright yellow-orange), swedenborgite (sky blue), and tilasite (orange), each against an orange-red background of brilliantly fluorescent calcite.

Red-fluorescent calcite in sizable, well-formed crystals is much less common than the massive material and is priced accordingly. Especially notable are the thick, platelike groups of large (1-3 inch) scalenohedral crystals from the Idarado mine near Telluride, San Miguel County, Colorado. The best of these are delicate pink in color, translucent, lustrous, sharply formed, and brightly fluorescent orange-red to pinkish red under the longwave lamp. Large, undamaged specimens are quite uncommon and command high prices driven even higher by the continuing mania for Colorado minerals. Equally attractive, and unbeatable for sheer form, are the complex groups of pale pink, chalky calcite crystals from the Pachapaqui mine, Huanuco Department, Peru. These too fluoresce brightest under the longwave lamp and occur in large, complex crystals, some with multiple terminations and others studded with smaller crystals, either alone or in elaborate groups aesthetically arranged on a contrasting matrix of black sphalerite. Localities in Mexico and South Africa have produced material of comparable attractiveness, and in recent years the mines at Tetjuche-Pristan, Primorskij Kraj, Russia have produced some outstanding specimens. This last locality has the potential for eclipsing all others; appearing in dealers' stocks recently have been some remarkable crystals, some more than eight inches across, well formed, and highly fluorescent.

"Terlingua-Type" Calcites

Most large museum displays of fluorescent minerals include a specimen of calcite from the mercury-mining district of Terlingua, Brewster County, Texas. Nearly all of this calcite is massive and of white to pale pink or pale green color in daylight; large blocks more than ten inches across have been found. More so than almost any other mineral, Terlingua calcite displays a remarkable transformation from drab and unappealing in daylight (some is downright ugly) to uncontestedly beautiful under ultraviolet radiation *(plates 2-4)*. Moreover, the fluorescence shows a remarkable color change from bright medium blue under the shortwave lamp to bright medium pink under the longwave lamp, and there remains after shortwave excitation a persistent blue phosphorescence whose intensity is matched by that of few other mineral species. Display-cabinet lamps fitted with timers, to cycle in succession from white light to longwave ultraviolet radiation, shortwave ultraviolet radiation, and then to darkness, are particularly effective in displaying the properties of this fascinating mineral. The cause of neither the blue nor the pink fluorescence is known, although the subject is currently under study.

Terlingua calcite displays an additional remarkable property: it slowly changes color in daylight. Specimens freshly broken generally are mottled white and pale bluish green, but the green areas, upon prolonged exposure to daylight, gradually fade and then turn pale pink. Specimens showing both colors—pink on the long-exposed display face but still pale bluish green on the base—are occasionally seen in old collections.

Calcite showing broadly similar fluorescence is known from several other localities, notably quarries in San Saba County, Texas, and the Sugar Creek quarry near Washington Court House, Fayette County, Ohio. Additional U.S. localities are mentioned in the 1983 and 1994 collectors' books by Manuel Robbins. In the late 1980s, however, perhaps the most remarkable of all Terlingua-type calcites made its debut on the mineral market by way of Mexico. This calcite occurs in large cleavage blocks, some of them almost perfectly transparent and of variable color: pale pink, yellow, brownish orange, dark green, colorless, and all shades in between. None of the yellow and little of the colorless calcite fluoresces, but nearly all that of pink, orange, or green color shows a good response. The shortwave fluorescence is much like that of Terlingua calcite except for the spooky effect caused by the utter clarity of the mineral: the blue glow has a pronounced ethereal quality, as if it were coming from a luminous hologram rather than a solid rock. The phosphorescence, too, is similar to that of Terlingua calcite but with the same strange, disembodied quality. Under the longwave lamp the response is variable and ranges in color from pink to yellow-orange, generally very bright and attractive. Claimed localities for this material seem to change with every new influx of specimens and now include five different states in Mexico, but the most recent and apparently reliable sources place the true locality near Múzquiz, a small town in Coahuila.

White- and Pastel-Fluorescent Calcite

Calcite that fluoresces white or various unsaturated (pale) hues of yellow, red, orange, green, and blue is known from hundreds of localities. The geologic environments in which such calcites are found are diverse and include hydrothermal veins, caves, spring and stream deposits, and vug and vein fillings deposited from ground water. The supposed activators are correspondingly diverse and include organic impurities, crystal defects, and trace amounts of rare-earth elements substituting for calcium. Relatively few natural examples have been sufficiently well studied that their activator systems are known. Moreover, measurement of their emission spectra may not be of much help in identifying activators because these calcites are "broad-band emitters," which is a short way of saying that the emitted light spans a broad range of wavelengths. An example is shown in figure 7 for a specimen of calcite, amber-colored in daylight, that occurs as small crystals lining vugs in limestone in the Faylor-Middlecreek quarry near Winfield, Union County, Pennsylvania. Here the emitted light spans nearly the entire visible range (roughly 360-750 nm for most humans). Whenever light of many wavelengths is seen by the

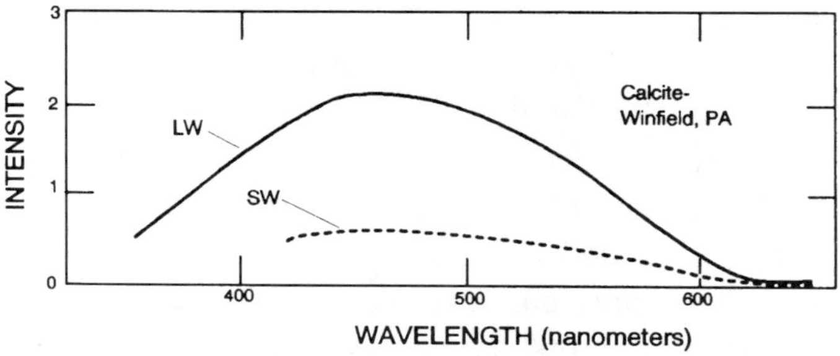

Figure 7. Fluorescence emission spectrum of calcite from the Faylor-Middlecreek quarry near Winfield, Pennsylvania (from Modreski, 1974, *Journal of the Fluorescent Mineral Society,* vol. 3, no. 1, p. 11-19). The emitted light spans a broad range of wavelengths and is perceived by our eyes as pale yellow to yellowish white. Solid line = longwave ultraviolet light; dashed line = shortwave ultraviolet light.

human eye, we see it as white or some other pale color, and in this instance we would describe the fluorescence as pale yellowish white. The geologic environment, daylight color of the fluorescent calcite, and its broad-band emission spectrum are consistent with an organic activator. Though these facts are by no means conclusive, it is worth noting that calcite showing pale-colored fluorescence is associated with organic matter at many localities, and at many more has formed in an environment where the former presence of organic matter is likely. Microscopic study of calcite from some of these localities reveals that tiny globules of oil became trapped within the crystals as they were growing and that the included oil, rather than the calcite itself, is responsible for the observed fluorescence. In other cases it is suspected that organic molecules incorporated within the calcite structure are the cause of fluorescence, but such calcites are notoriously difficult to study. Only this decade, for example, was it shown that the cause of the characteristic pale greenish- to bluish-white fluorescence of much cave calcite worldwide is due to fulvic and humic acids, complex organic acids derived from decaying vegetation and carried in solution by groundwater into the caves where the calcite grew.

Fluorescence in many other calcites is suspected to be activated by defect structures. As evidence we may consider two results from a 1975 study of nearly 350 fluorescent calcite samples: (1) the impurity content of the white-fluorescent calcites examined was so variable that likely impurity activators could not be identified, and (2) some of the samples analyzed were very nearly pure but nevertheless fluoresced like the others. For the pure samples, especially, activation by crystal-structure defects was strongly suspected. Moreover, the cloudy appearance of some of the samples in daylight is consistent with the presence of abundant structural defects, and much of the calcite was inferred from geologic evidence to have grown rapidly from low-temperature solutions, conditions particularly conducive to defect formation. These facts, however, constitute only indirect evidence of defect-activated fluorescence and are hardly conclusive; low-temperature conditions of formation are exactly those that favor the presence of organic activators as well. Proof of defect-activated fluorescence and demonstration of the activator mechanism require study with sensitive and sophisticated instruments available to relatively few scientists.

The third group of activators mentioned above, the rare-earth elements, is also a known cause of white or near-white fluorescence in calcite. The presence of rare-earth impurities in some fluorescent calcites is evident from their emission spectra, which, though broad, typically contain multiple sharp peaks. Not surprisingly, chemical analysis of many calcites showing such multiple-peaked spectra disclosed rare-earth contents 10-100 times greater than those of other calcite samples whose pale-colored fluorescence was instead suspected of being due to defect activators.

The attractiveness of pale-fluorescing calcite is purely a matter of personal taste. Though some people prefer bolder colors, such as those shown by the legendary fluorescent minerals from Franklin and Sterling Hill, the charm of a well-formed calcite crystal softly glowing pale yellow or pure white can hardly be denied. Even massive specimens can have considerable appeal if calcite showing varied hues of fluorescence is intergrown in texturally interesting ways. The massive calcite of Ilwaco, Pacific County, Washington, is particularly notable in this regard. Once collected in abundance and still seen occasionally on the market today, this calcite shows a beautiful range of fluorescent colors—white, yellow, orange, pale blue—often in the same piece. The locality is now a state park and is off limits to collecting.

Green-Fluorescent Calcite

Calcite showing bright yellow-green fluorescence is fairly common; you will meet with it in many mines and in some caves and quarries. (Note, however, that collecting *anything* in caves is strongly discouraged in the United States, where the guiding maxim is "Take only photographs and leave only footprints".) Attractive examples of green-fluorescent calcite are known from Bisbee, Arizona, where delicate, diverging crystal groups have formed in caves above the orebodies. Similar calcite has been reported from several mining districts in New Mexico and Mexico, from Colorado, and from more than two dozen countries worldwide. The color of fluorescence, nearly identical to that of many secondary uranium minerals, has led many to assume that the response is due to uranyl ions as an impurity constituent of the calcite. Recent investigation of green-fluorescing calcite from several localities, however, has shown that the green response is due to microscopic films of hyalite opal intergrown with or encrusting the crystals, not to the calcite itself. The same is true of the related mineral aragonite. A photograph of a portion of one such encrusted crystal, highly magnified, is shown in figure 8. The coatings commonly are too thin and individual opal globules too small to be detected with a hand lens, but the patchy nature of many coat-

Figure 8. Patchy coating of globular opal on blades of aragonite from a cave in Clear Creek Canyon, Jefferson County, Colorado. The photograph is a highly magnified view as seen with a scanning electron microscope. The aragonite crystals are very small and the opal films on them nearly invisible under a hand lens; the scale bar at bottom left is only 0.1 mm long. Photograph by P. J. Modreski, U.S. Geological Survey.

ings as seen under an ultraviolet lamp is a good clue to their presence.

Aragonite

Aragonite is a mineral well known to mineral enthusiasts for its variety of forms. Some of it occurs as large, lustrous, and sometimes colorful crystals much sought by collectors of aesthetic mineral specimens. The "decorations" in many caves worldwide—including much flowstone and dripstone, and also the radiating clusters of elongate crystals known as anthodites, or "cave flowers"—commonly are aragonite, or aragonite interlayered with calcite. Aragonite is a principal component of sea shells and coral, and thus of the beach sands and limestones derived from them. It is also the principal mineral component of pearls. Massive aragonite also occurs as a vein-filling mineral; such veins commonly exhibit an intricately layered structure due to sequential deposition from the walls of an open fracture toward the center. Aragonite has the same chemical formula as calcite but is much the less common of the two for two reasons: (1) it forms in nature under a much narrower range of conditions, and (2) with time, it inverts to calcite through a change in internal atomic structure. Because of this last factor there are no geologically old aragonites, though many specimens that are now calcite preserve perfectly the visual appearance of the original aragonite.

Much aragonite fluoresces, some of it spectacularly. Only a few examples from well-known localities are mentioned below to illustrate some of the collecting possibilities offered by this remarkable mineral.

Sicily, Italy

Crystal specimens of aragonite from the sulfur mines of southern Sicily provide the standard against which most other aragonite is measured, both for beauty in daylight and for fluorescence. Specimens from old collections commonly are labeled Girgenti, a province name that corresponds to present-day Agrigento. Much Sicilian aragonite ranges in daylight color from pale grayish green to creamy white and forms pseudohexagonal crystals 0.5-2 inches across and 1-4 inches long, some in impressively large groups attractively associated with yellow sulfur. The adjective "pseudohexagonal", incidentally, is used here advisedly because of one of Nature's little tricks: crystals of aragonite commonly assume the shape of a hexagonal column even though aragonite crystallizes in the orthorhombic system, within which true hexagonal symmetry is not possible. The hexagonal form is the result of repeated cyclic twinning, common in aragonite and a few other minerals.

The fluorescence of these aragonite groups is variable but nearly always bright and beautiful. Under the longwave lamp the response ranges from deep pink through bluish pink to pinkish blue. Specimens fluorescing "hot" pink are much in demand and offer a color duplicated by few other fluorescent minerals. This fascinating and complex fluorescence, coupled with appealing crystal form, make fine specimens of Sicilian aragonite among the most dramatic ornaments of any fluorescent display. Moreover, many of the crystals are markedly phosphorescent and show an enduring pale green glow when the longwave lamp is extinguished. The contrast in color between the fluorescence and phosphorescence is among the most extreme in the mineral kingdom. As if all this were not enough, Sicilian aragonite fluoresces brightly under the shortwave lamp as well—but now in shades of white rather than pink or blue. The white response is not as appealing to most eyes as the pink, but an attractive textural element is added because growth zones in many crystals are highlighted as concentric hexagons of differing fluorescent brightness on the flat terminations of the columnar crystals. This effect generally is less evident under longwave ultraviolet radiation.

Friedensville, Lehigh County, Pennsylvania

The Friedensville zinc mine near Bethlehem, Pennsylvania, once produced dull-lustered masses of cream-colored aragonite showing an interesting bicolored fluorescence: bright golden yellow under the longwave lamp, but bright white tinged with blue under the shortwave lamp. There remains after exposure to either lamp a pale blue phosphorescence of fairly long duration. The fluorescence in many specimens is not distributed evenly throughout the mass but is patchy, adding to the overall visual interest. Analyses show a few percent zinc in this material, but it has not (to my knowledge) been proven that the zinc has anything to do with the peculiar response. Though properly termed *zincian aragonite,* many specimens in old collections are labeled with the informal varietal name "nicholsonite". The name, originally applied to zincian aragonite from the Leadville mining district in Colorado, honors Mr. S.D. Nicholson, former Chief Geologist of the Western Mining Company.

Santa Eulalia Mining District, Chihuahua, Mexico

The mines of the Santa Eulalia district in north-central Mexico have long produced fine mineral specimens and are justifiably famous as one of the world's premiere mineral localities. Among the species recovered in quantity is aragonite, nearly all of it from caves near the orebodies, much as at Bisbee, Arizona. The association of caves with the sulfide ores at both localities is hardly accidental: alteration of the original ore minerals by seeping ground water rendered the water highly acidic, which accelerated dissolution of the limestone host rocks to form the caves. The same alteration resulted in a net reduction in volume of the ore, creating additional void space. The caves at Santa Eulalia are particularly large and festooned with fantastic arrays of crystals encrusting the walls, coating cave floors, and hanging from ceilings. Large crystal groups of aragonite, and of aragonite intergrown with calcite, have long been standard items on the mineral market. For years they have been so plentiful that many are sold by weight rather than by the piece, and prices are low. The very best specimens, however—those combining attractive shape, clean white color, and lack of damage—are separated for sale individually and command higher prices. Some of them are shaped like cauliflowers and are sold accordingly as "cauliflower aragonite;" others are delicate groups of long, tapering crystals. Their fluorescence is typical of cave aragonite: moderately bright white to creamy white under the longwave lamp, followed by a white phosphorescence tinged with green; and moderately bright bluish white under the shortwave lamp, followed by a bluish-white to pale blue phosphorescence. The difference in color of fluorescence under longwave and shortwave ultraviolet light is slight but readily noticeable, and it is typical not only of aragonite from many localities but of other species as well, including some strontianite, calcite, celestite, and gypsum. The response in aragonite from some caves is so readily induced that a flashlight beamed across a section of cave wall and then extinguished will leave behind a faint but persistent phosphorescent trail.

Adding to the attractiveness of some aragonite specimens are patchy coatings of hyalite opal that fluoresce bright yellow-green under the shortwave lamp. Specimens from Santa Eulalia are particularly distinctive, for often it is only the knobby tips of individual branches of aragonite that are coated with hyalite.

Other Carbonates

Smithsonite, magnesite, dolomite, cerussite, strontianite, and witherite are other carbonate species commonly reported as fluorescent and found in many systematic collections. Witherite is perhaps the most coveted of these. Like aragonite, it commonly forms twinned crystals of pseudohexagonal form, many of them large and with horizontally striated or corrugated surfaces. Some witherite fluoresces brilliant bluish white, more strongly under longwave than shortwave ultraviolet light; handsome groups add considerably to the appeal of any fluorescent display. The most notable localities for fine, brightly fluorescent crystal groups include several mines in England and the area around Rosiclare, Illinois. Cerussite, too, occasionally can be found in handsome crystal groups showing good fluorescence, in this case generally pale to medium yellow, though it should be noted that the response of many specimens is not quite bright enough for effective display. Cerussite is a widespread mineral, common in the upper reaches of many mines where lead-containing minerals are present in the ore. Diligent search by a collector may occasionally be rewarded with attractive specimens, but cerussite and the other carbonates mentioned above are most likely to be found in specimens of greater geologic than aesthetic interest. Exceptions include the lovely red-fluorescent smithsonite from the Empire mine in the Fiero-Hanover district, Grant County, New Mexico, and the bright white-fluorescent crystal groups of strontianite from western Germany. These are well worth the sometimes considerable effort required to obtain them. The hunt for still rarer species such as red-fluorescent otavite and norsethite can take years if fine examples are the objective. The dedicated collector, however, sees this not as a difficulty but a challenge; the quest for fine minerals is often fully as enjoyable and rewarding as possessing them.

Uranium Minerals

The uranium minerals include species from several different mineral groups but are here considered together for convenience because many are fluorescent and are activated by the uranyl ion, an essential part of their chemical makeup. The list of uranium minerals is a long one—there are more than 170 of them—and their ranks include phosphates, arsenates, silicates, carbonates, molybdates, sulfates, oxides, selenates, and vanadates of numerous metals in various combinations and degrees of hydration. The uranium boom of the 1950s and 1960s saw an unprecedented surge in the study of uranium minerals, many of which previously were unknown to science, and some of which had been incompletely or wrongly described, or even mistaken for other species. The huge task of sorting out the uranium minerals continues even today; it sometimes seems as if there is no end to the strange rare-earth uranium species from the mines of Zaire, and several new species have been described recently from long-abandoned mines in the western United States. The rarity of many species, the tendency of many of them to look alike, their complex chemistry, and common intergrowths of two or more species are some of the reasons the study of uranium minerals is so very challenging.

Uranium minerals can be divided into two large groups,

those which contain tetravalent uranium (the U^{4+} ion) and those which contain hexavalent uranium (U^{6+}); the latter, in combination with two oxygen ions, forms the uranyl ion (UO_2^{2+}). Minerals of the former group include uraninite and coffinite, the major uranium ore minerals of the western United States, but these do not fluoresce. Visitors to the uranium mines of this region often are disappointed to find only black, fine-grained, nondescript uranium minerals disseminated in sandstone and showing no response to the ultraviolet lamp. In the near-surface, oxidized portions of some of these same deposits, however, some of the secondary uranium species—those containing hexavalent uranium in the form of uranyl ions—fluoresce brightly and provide an obvious clue to the underlying orebody. The portable, battery-operated ultraviolet lamp, in combination with the venerable Geiger counter, was responsible for one of the strangest periods in U.S. mining history, when hundreds of prospectors in search of riches swarmed over the desert Southwest both day and night. Some were successful, but their search was hampered somewhat by a quirk of geology: two of the most common and abundant of the secondary uranium species, carnotite and tyuyamunite (hydrous uranyl vanadates of potassium and calcium, respectively), generally do not fluoresce.

Those uranium minerals which *do* fluoresce number more than 80 and include some of the most brightly fluorescent species known. Autunite and its dehydration product meta-autunite are the most common of these and have been reported as fluorescent from more than 130 localities. The large plates of thickly grouped autunite crystals from the Daybreak mine (now closed) on Mt. Spokane, Spokane County, Washington, are among the world's finest; their brilliant yellow-green fluorescence *(plate 8)* is unforgettable. Many other uranium minerals fluoresce a similar color, some just as brightly and others not. Colors of fluorescence of uranium minerals as a group, however, are somewhat more variable than is generally realized, and a display of many of them together can be most instructive. The glow of liebigite and andersonite, for example, is distinctly green to bluish green, much like that of a traffic light and clearly different from the yellow-green of autunite. A few other uranium species, though rare, fluoresce pale yellow to deep orange. The cause of these latter colors remains unknown.

With the passing of the uranium boom, uranium minerals have become less common on the collector market than they used to be. Many U.S. deposits have long been mined out, or abandoned because of low prices. Still, there are so many known localities for uranium minerals that finding some for sale and assembling a small collection should not be difficult. Should you decide to collect uranium minerals, please exercise some caution: they *are* radioactive, some much more so than others, so take care to avoid direct exposure to your specimens as much as possible. Moreover, uranium is a toxic metal when ingested. Some people keep their specimens in lead-lined boxes or small closets to reduce radiation leakage. Storage in a well-ventilated area is important because uranium minerals generate radon, an odorless, radioactive gas readily absorbed by dust particles in the lungs, where it poses a distinct health hazard (particularly to smokers). Placement of specimens—especially crumbly ones—in sealed plastic bags will minimize chances of both dust inhalation and leakage of radon into the storage area. The best solution is to keep the specimens in some out-of-the-way area removed from your living quarters but still accessible for study.

Gemstones

The gemstones constitute a mineral class defined not by science but by commerce: a gemstone is any mineral or rock that, when cut and polished, possesses sufficient beauty and durability to be used as a personal adornment or other ornament. In practice the definition is extended to embrace other materials as well, including organic substances such as amber, fossils such as corals and ammonites that show an attractive pattern or color when polished, and synthetic materials made in the laboratory. Note that durability is essential to the definition of gemstone and separates the true gem minerals from those others that are merely beautiful. Some carbonate minerals, for example—notably rhodochrosite, calcite, smithsonite, sphaerocobaltite, and cerussite—occur in richly colored crystals of sufficient size and transparency to be cut into fine gems, but all of them combine easy cleavage with low hardness and hence are too readily damaged to have any practical value for jewelry.

Already beautiful in daylight, many gemstones take on another, totally different kind of beauty under the ultraviolet lamp. A collection of rough stones as they are taken from the ground, together with finished articles of jewelry containing the same gem species in polished form, can be a particularly fascinating one, and for many gemstones not nearly as expensive as one might imagine. The alphabetical list of fluorescent gemstones given below is only a sampler to indicate some of the possibilities open to collectors in this most elegant of specialties.

Amber

Amber, the fossilized resin from coniferous trees, is much prized for its warm yellow through brown color and high luster when polished. Polished lumps of amber often are set in brooches or strung into bead necklaces; larger lumps sometimes are used for carving. Amber by its very nature occurs most plentifully in localities that offer few other possibilities for fluorescent minerals: in coal mines, clay and sand pits, and, in some regions, washed up on beaches. Many dozens of localities are known, though today most commercial amber comes from two main sources: the Baltic area, which stretches along the northern coast of Poland eastward to the Kaliningrad Peninsula, and the Dominican Republic. Most amber fluoresces, generally bluish white to medium blue but also yellow, cream, and brown, in some pieces tinged with green. The brightness of fluorescence varies over a wide range, especially from one locality to another, but almost always it is strongest under the longwave lamp. Some of the finest Dominican amber, the type referred to informally as *blue amber*, owes its name to a fluorescence so bright that in sunlight the surface of the stone appears as if suffused with a blue haze. The lovely pale yellow amber recently introduced on the market from Colombia, in contrast, generally shows only a dim response. Baltic amber, still so plentiful that one merchant at the annual Tucson Gem and Mineral Show habitually offers an entire steamer trunk full of large chunks, shows as much variability under the lamp as it does in daylight. One can spend many a fascinating hour sorting through specimens of amber under the lamp; those showing rich color and attractive texture make fine display pieces.

Beryl

Beryl, though an important gem species, almost never is sold as such in the gem trade; instead it is marketed under a series of varietal names based principally on color. *Emerald,* the deep-green gem beryl, is the most well-known and coveted variety, but there are others: *aquamarine* (pale to medium blue through bluish green), *goshenite* (colorless), *heliodor* (lemon yellow through gold), *bixbite* (strawberry red, not to be confused with the unrelated mineral species *bixbyite),* and *morganite* (pale to medium pink, rose, or violet). As a fluorescent mineral beryl is uncommon; the 1989 *Journal of the Fluorescent Mineral Society* lists only 40 reports of fluorescence in the species, a small total relative to the very large number of localities at which the mineral occurs. Various hues of green are the most commonly mentioned colors of fluorescence, but beryl specimens fluorescing red, yellow through orange, and bluish white have been reported. Intensities range from weak to moderate and are generally strongest under the shortwave lamp, but it must be admitted that most gem beryls are far more appealing in daylight than under ultraviolet light.

Activators of fluorescence in beryl are poorly understood except for chromium, whose effects as an impurity constituent are fascinating: chromium not only imparts a deep green color in daylight to the finest emeralds but also is responsible for the weak, deep *red* fluorescence of some natural and synthetic stones. Colombian emeralds, generally low in iron and hence showing more readily visible fluorescence than emeralds from most other localities, are the most notable of these. Green-fluorescing emeralds have also been reported, but the activator is unknown.

Corundum

Corundum, like beryl, generally is marketed under other names. *Ruby*, the red gem corundum, in fine stones is among the most valuable of all gems, more so than even emerald and diamond. *Sapphire* is a catch-all term for all other gem corundum. Though we most commonly think of sapphire as blue, it can also be colorless, yellow, orange, green, purple, pink, and shades of red too pale to justify the term ruby.

The deep red fluorescence of a fine specimen of ruby under the longwave lamp is unforgettable. Its pure red light has been described by some authors as resembling a glowing coal or smoldering ember, so much so that people unfamiliar with the nature of this light may refuse to touch the stone for fear of being burned. As with emerald, the red fluorescence is due to trace amounts of chromium. The fluorescence is weakest under the shortwave lamp, generally much stronger under the long-wave lamp, and stronger still when the stone is placed under visible blue light. This latter effect, of fluorescence in one part of the visible spectrum being excited by light in another part of the visible spectrum, is rather common among the chromium-containing gems. It has long been the basis of a standard gemological test to distinguish rubies from other gem species of similar appearance, notably some members of the garnet and tourmaline groups, by viewing the stone under a beam of blue light generated by passing sunlight through a dilute solution of copper sulfate. The test is not infallible, however, for genuine rubies containing too much iron impurity will fluoresce weakly, and from some localities not at all.

Purple, pink, and pale red varieties of corundum commonly show a red fluorescence akin to that of ruby and due to the same cause. Fine blue sapphires generally show no notable fluorescence, though some show a faint red glow under the long-wave lamp (again, due to a trace of chromium), and others show a weak blue to bluish-white response under the short-wave lamp. Among the most attractive fluorescent sapphires are the colorless to pale yellow crystals from the gem gravels of Sri Lanka (formerly Ceylon), many of which show an attractive yellow-orange fluorescence, in some crystals bright enough for display. Most of the crystals range in length from one-half to one inch, but some are as much as six inches long, and nearly all are sharply formed, with lustrous faces in no need of polishing to appreciate the beauty of the stone. A small collection of such crystals, showing mixed red and orange fluorescence under the longwave lamp, is a pleasing addition to any fluorescent display.

Diamond

Diamond, one of several minerals composed of nearly pure elemental carbon, undoubtedly is the gemstone most familiar to the general public. It is considerably less well known as a fluorescent mineral, even among collectors, perhaps because most people still believe diamonds to be rare and invariably expensive. However, if good fluorescence rather than a flawless stone is your goal, you should find it possible to obtain industrial-grade diamonds at a per-carat cost of less than you would pay for a decent dinner at a modest restaurant. Though most gem merchants carry only finished stones, some carry rough specimen diamonds as well; major gem and mineral shows are good places to look, and some dealers advertise in well-known hobbyist and trade journals.

The percentage of natural diamonds that show visible fluorescence is variable from one locality to another. Estimates of the worldwide average range from as low as 10-15% to 50% or more, the higher estimates seeming to gain favor over time. The great majority of diamonds fluoresce much more strongly under longwave than shortwave ultraviolet light, though exceptions are known. Many, too, are strongly phosphorescent, some even after only brief exposure to sunlight. The enduring phosphorescence of diamond has been known for centuries and is part of the historic allure of the species, which to early scientists seemed to possess magical powers unknown in other gemstones. The most common colors of fluorescence in diamond are medium blue and pale to medium yellowish green, but stones fluorescing yellow, orange, reddish orange, orange-red, and white are also known *(plate 1)*. The colors of phosphorescence are likewise variable and in many stones differ markedly from that of the light emitted while the lamp is still on. Stones fluorescing intense sky blue yet phosphorescing bright yellow to yellowish orange are common from some localities and offer one of the most extreme examples known of color change in a fluorescent mineral species.

The variety of colors exhibited by fluorescing diamonds is indicative of a complex variety of activators, both of the impurity and defect type. Known activators include nitrogen and aluminum substituting together for carbon atoms (blue fluorescence), single nitrogen atoms coupled with a carbon vacancy (yellow fluorescence), two nitrogen atoms associated with crystal defects due to radiation damage (yellowish-green fluorescence), and segregations of submicroscopic platelets of nitrogen along selected crystallographic planes within the diamond structure (greenish-yellow fluorescence). The relative intensity of the fluorescent light emitted by these activators, in diamonds from different areas, varies with their geologic conditions of formation and with their history of exposure to radioactivity.

Spinel

Spinel is a popular gem species that exhibits a wide range of colors in daylight: all shades of red, deep blue through violet-blue, deep green, and in rare stones almost colorless. Most valuable are gem spinels of deep red color, which sufficiently resemble ruby that the two are commonly confused. The historic Black Prince's Ruby, the large, deep red stone set front-center in the Imperial State Crown of the British Royal Family, is one such stone; it is not a ruby as once supposed, but a magnificent uncut spinel. In modern times, many of the finest spinels have come from the Mogok Stone Tract in Myanmar (formerly Burma), where they occur as deep red, transparent, sharply formed octahedral crystals of brilliant luster. The fluorescence of much red spinel, like that of ruby, is activated by chromium. Under the longwave lamp the two species may appear identical to the eye, but their fluorescence spectra are sufficiently different that they may readily be distinguished by viewing the emitted light through a hand spectroscope. Pink and violet stones commonly show a similar red fluorescence, but in many stones the response is weak or absent because of the quenching effect of small amounts of iron.

Fluorescence in spinels of blue to violet daylight color is not nearly as common as in those of red hue. Reported colors of fluorescence, in addition to the familiar chromium-induced red, include green, orange, and blue. The response typically is strongest under the longwave lamp and is much weaker, or absent, under the shortwave lamp.

Topaz

Topaz is another of the multihued gemstones that have found widespread popularity with the public. Colorless and blue stones are most common, but natural green, brown, orange, and pink topazes are also known. The lovely sherry yellow to orange topaz from Ouro Preto, Brazil is particularly prized and is often marketed as "Imperial topaz". Fluorescence in topaz is hardly rare, though good crystal specimens showing a bright response are known from few localities and have never been plentiful on the mineral market. Attractive, blocky crystals showing bright golden-yellow fluorescence under the longwave lamp have been recovered from several localities, notably the Little Three mine in the Ramona pegmatite district, San Diego County, California, and from pockets in granitic igneous rocks of the Sawtooth Range, Idaho. Topaz fluorescing bluish white, pale green, or white under the shortwave lamp is also known, and a red to pink response under the longwave lamp has also been described. This last response is activated by chromium but is only weakly excited by conventional longwave ultraviolet lamps; visible light of blue or yellow to green color is much more effective in producing it. Other than chromium, the activators of fluorescence in topaz remain unknown despite recent chemical study.

Zircon

Zircon is a minor mineral component of many igneous rocks and occurs in an impressive range of colors. Yellow, orange, red, brownish red through brown, and various shades of green have been reported for natural stones, and blue, colorless, and golden brown for heat-treated ones. Regardless of daylight color, the characteristic color of fluorescence for this species is some shade of yellow, commonly golden yellow, in some specimens with an orange tint. The intensity of fluorescence varies dramatically from one locality to another and even within different parts of the same crystal. Some zircon fluoresces brightly, some feebly, and some not at all, but the response generally is strongest under the shortwave lamp. The activator of yellow fluorescence in zircon is trivalent dysprosium, one of the rare-earth elements most commonly present in zircon as a substitutional impurity. The slight color variations probably are due to the differing proportions of other rare-earth elements present.

Zircon is known from so many localities that specimens are fairly common on the mineral market, and cut stones are plentiful in the gem trade. Perhaps most familiar to mineral collectors are the large (½-2 inch long), deep reddish-brown, sharply formed crystals embedded in black micaceous matrix from Seiland Island, Finnmark, Norway; these fluoresce bright golden yellow under the shortwave lamp and in well-formed groups make striking specimens for display. Zircon in brightly fluorescent masses has also been produced in moderate abundance from near Mud Tank, Northern Territory, Australia; some specimens are as large as a baseball and are fragments of still larger crystals. Zircon is also a component of much beach sand and stream sediments worldwide and in many areas can be detected instantly by means of a portable ultraviolet lamp. A concentrate of heavy minerals rich in tiny, fluorescent zircon grains can often be obtained by panning the sand, much as one would for gold.

Other Gemstones

Many other gemstone species besides those few described

here are known to fluoresce, including (but hardly limited to) andalusite, benitoite, chrysoberyl, clinohumite, danburite, elbaite, fluorapatite, jadeite, opal, petalite, quartz (especially the variety chalcedony), scheelite, spodumene, taaffeite, and vesuvianite. Some of these species are more likely to be found in the stock of dealers in specialty gems than as specimens on the mineral market. Collectors interested in assembling a comprehensive collection of fluorescent species would thus be well advised to frequent gem merchants as well as mineral dealers for specimens. Though faceted gems of many species command premium prices, commonly the rough, uncut material is considerably less expensive and well within the reach of many collectors. Remember, too, that tiny internal flaws and small differences in color and transparency—factors of much greater importance to the collector of gems than to the collector of fluorescent minerals—can make an enormous difference in price. Prize fluorescent specimens can sometimes be found among the lower grades of gem rough and purchased at economic rates. Before assembling your collection of fluorescent gemstones, however, be sure to read at least one comprehensive book on the subject so that you will know something about identification methods and gemstone treatment. Many gemstones on the market today have been altered in some way, commonly through heating or irradiation to improve their color, impregnation by oils or glues to improve their clarity and durability, etc. Though such treatment is a well known and accepted part of the gem trade, the fluorescence of the parent material is in some cases altered, and the collector should be aware of this.

Further Reading

A short bibliography is offered here for those readers who wish to pursue additional knowledge through other sources. Included are references to some of the principal hobbyist publications of past decades, as well as to selected modern publications of special interest. For those wishing a fuller treatment, the extensive reference lists in the two books by Manuel Robbins are particularly recommended as entries to the vast semitechnical and technical literature on mineral fluorescence.

Dake, H.C., 1953, *The uranium and fluorescent minerals:* The Mineralogist Publishing Company, Portland, Oregon, 79 p. [An updated version of DeMent's original report cited below, with additional chapters on prospecting methods for uranium]

DeMent, Jack, 1949, *Handbook of fluorescent gems and minerals:* The Mineralogist Publishing Company, Portland, Oregon, 68 p. [Brief descriptions of the fluorescence of numerous mineral species, with a chapter devoted to gemstones]

Gleason, Sterling, 1960, *Ultraviolet guide to minerals:* D. Van Nostrand Company, Inc., Princeton, New Jersey, 244 p. [For many years the most popular work for the hobbyist; now much outdated but still enjoyable]

Harvey, E.N., 1957, *A history of luminescence from the earliest times until 1900:* Memoirs of the American Philosophical Society, vol. 44, 692 p. [A delightful book that offers 69 pages (!) of references on the title subject]

Jones, R.W., Jr., 1964, *Nature's hidden rainbows—the fluorescent minerals of Franklin, New Jersey:* Ultra-Violet

Products, Inc., San Gabriel, California, 120 p. [Popular account of the fluorescent minerals from the world's most prolific locality, summarizing the state of knowledge as it existed 30 years ago]

Journal of the Fluorescent Mineral Society, 1972-present. [The Fluorescent Mineral Society, founded in 1972, is an international organization of about 300 members devoted to disseminating information about mineral fluorescence. The society publishes an annual to biennial journal containing nontechnical to semitechnical papers; it also publishes a bimonthly newsletter. Those interested in purchasing past issues or in joining the society should write: Fluorescent Mineral Society, P.O. Box 572694, Tarzana, California 91357, U.S.A.]

Palache, Charles, 1935, *The minerals of Franklin and Sterling Hill, Sussex County, New Jersey:* U.S. Geological Survey Professional Paper 180, 135 p. [Still the definitive work on the minerals of these fascinating deposits. Includes much descriptive information on each species then known, with comments on fluorescence]

Robbins, Manuel, 1983, *The collector's book of fluorescent minerals:* Van Nostrand Reinhold, New York, 289 p. [Ambitious effort to explain the fluorescence of hundreds of mineral species to the hobbyist. Includes much information on human perception of color]

Robbins, Manuel, 1994, *Fluorescence—gems and minerals under ultraviolet light:* Geoscience Press, Inc., Phoenix, Arizona, 374 p. [A good, basic reference for hobbyists, with information on the fluorescence of more than 460 mineral species. Extensive bibliography]

Schumann, Walter, 1977, *Gemstones of the world* [English translation by Evelyne Stern of original German text]: Sterling Publishing Company, Inc., New York, 256 p. [A fine general-purpose reference book for nonspecialists. Includes many color photographs of both rough and cut gems and a tabulation of physical properties, including fluorescence, for each gem mineral species]

Sinkankas, John, 1964, *Mineralogy for amateurs:* D. Van Nostrand Company, Inc., Princeton, New Jersey, 585 p. [Highly readable and informative; one of the best introductory texts on mineralogy]

The Picking Table, 1960-present. [*The Picking Table* is the twice-yearly journal of the Franklin-Ogdensburg Mineralogical Society, Inc. Papers in *The Picking Table* range from nontechnical to semitechnical and contain information on all apects of the mineralogy and mining history of the two zinc ore bodies at Franklin and Sterling Hill, New Jersey, source of many of the world's finest specimens of fluorescent minerals. The society has been active since 1959. Information on fluorescence, though not a focus of the journal, is contained in numerous issues. People interested in purchasing past issues or in joining the society should write: Franklin-Ogdensburg Mineralogical Society, Inc., Box 146, Franklin, New Jersey 07416, U.S.A.]

Verbeek, E.R., and Modreski, P.J. (eds.), 1989, *The Henkel glossary of fluorescent minerals:* Journal of the Fluorescent Mineral Society, Inc., vol. 15, 91 p. [The largest published compendium of the colors of fluorescence of minerals; includes 566 mineral species, 21 organic substances, and 38 composite materials such as rocks]

Webster, Robert, 1975, *Gems—their sources, descriptions and identification (3rd ed.):* Newnes-Butterworths, London, 931 p. [A standard and well-respected text in the field of gemology. Includes general descriptions of the fluorescence of numerous gem species plus a chapter on fluorescence testing of gemstones]

Table 1.
The Chemical Elements and Their Symbols

1. Symbols are listed in alphabetical order to facilitate reading of chemical formulas given in the text.

2. Elements not found in minerals are not listed. These include (a) short-lived radioactive elements such as protactinium and technetium and (b) the transuranium elements, which do not occur naturally on Earth but have been prepared experimentally during nuclear reactions. The noble gases (helium, neon, argon, krypton, xenon, radon) are retained in the list but generally are absent from minerals; a few of them are widespread in minute amounts as a result of natural radioactive decay.

Ag	Silver	Hf	Hafnium	Re	Rhenium
Al	Aluminum	Hg	Mercury	Rh	Rhodium
Ar	Argon	Ho	Holmium	Rn	Radon
As	Arsenic	I	Iodine	Ru	Ruthenium
Au	Gold	In	Indium	S	Sulfur
B	Boron	Ir	Iridium	Sb	Antimony
Ba	Barium	K	Potassium	Sc	Scandium
Be	Beryllium	Kr	Krypton	Se	Selenium
Bi	Bismuth	La	Lanthanum	Si	Silicon
Br	Bromine	Li	Lithium	Sm	Samarium
C	Carbon	Lu	Lutetium	Sn	Tin
Ca	Calcium	Mg	Magnesium	Sr	Strontium
Cb	Columbium*	Mn	Manganese	Ta	Tantalum
Cd	Cadmium	Mo	Molybdenum	Tb	Terbium
Ce	Cerium	N	Nitrogen	Te	Tellurium
Cl	Chlorine	Na	Sodium	Th	Thorium
Co	Cobalt	Nb	Niobium*	Ti	Titanium
Cr	Chromium	Nd	Neodymium	Tl	Thallium
Cs	Cesium	Ne	Neon	Tm	Thulium
Cu	Copper	Ni	Nickel	U	Uranium
Dy	Dysprosium	O	Oxygen	V	Vanadium
Er	Erbium	Os	Osmium	W	Tungsten
Eu	Europium	P	Phosphorus	Xe	Xenon
F	Fluorine	Pb	Lead	Y	Yttrium
Fe	Iron	Pd	Palladium	Yb	Ytterbium
Ga	Gallium	Pr	Praseodymium	Zn	Zinc
Gd	Gadolinium	Pt	Platinum	Zr	Zirconium
Ge	Germanium	Ra	Radium		
H	Hydrogen	Rb	Rubidium		

*Niobium (Nb) and Columbium (Cb) refer to the same element. In most references the former is used.

Table 2.
Some Ion Groups in Fluorescent Minerals

Ammonium	$(NH_4)^{1+}$	Molybdate	$(MoO_4)^{2-}$
Arsenate	$(AsO_4)^{3-}$	Nitrate	$(NO_3)^{1-}$
Bicarbonate	$(HCO_3)^{1-}$	Oxalate	$(C_2O_4)^{2-}$
Borate	$(BO_3)^{5-}$	Phosphate	$(PO_4)^{3-}$
Carbonate	$(CO_3)^{2-}$	Selenate	$(SeO_4)^{2-}$
Chromate	$(CrO_4)^{2-}$	Silicate	$(SiO_4)^{4-}$, $(Si_2O_7)^{6-}$, etc.
Disulfide	S_2^{2-}	Sulfate	$(SO_4)^{2-}$
Hydronium	$(H_3O)^{1+}$	Tungstate	$(WO_4)^{2-}$
Hydroxyl	$(OH)^{1-}$	Uranyl	$(UO_2)^{2+}$
Mellate	$[C_6(COO)_6]^{6-}$	Vanadate	$(VO_4)^{3-}$

Table 3.
Chemical Formulas for Minerals Discussed in Text

Adamite, $Zn_2(AsO_4)(OH)$
Agrellite, $NaCa_2Si_4O_{10}F$
Andalusite, Al_2SiO_5
Andersonite, $Na_2Ca(UO_2)(CO_3)_3 \cdot 6H_2O$
Anglesite, $PbSO_4$
Anhydrite, $CaSO_4$
Ankerite, $Ca(Fe^{2+},Mg,Mn)(CO_3)_2$
Aragonite, $CaCO_3$
Autunite, $Ca(UO_2)_2(PO_4)_2 \cdot 10\text{-}12H_2O$
Barite, $BaSO_4$
Barytocalcite, $BaCa(CO_3)_2$
Benitoite, $BaTiSi_3O_9$
Benstonite, $(Ba,Sr)_6(Ca,Mn)_6Mg(CO_3)_{13}$
Beryl, $Be_3Al_2Si_6O_{18}$
Calomel, Hg_2Cl_2
Calcite, $CaCO_3$
Cancrinite, $Na_6Ca_2Al_6Si_6O_{24}(CO_3)_2$
Carbonate-fluorapatite, $Ca_5(PO_4,CO_3)_3F$
Carbonate-hydroxylapatite, $Ca_5(PO_4,CO_3)_3(OH)$
Celestite, $SrSO_4$
Cerussite, $PbCO_3$
Chalcopyrite, $CuFeS_2$
Chlorapatite, $Ca_5(PO_4)_3Cl$
Chondrodite, $(Mg,Fe^{2+})_5(SiO_4)_2(F,OH)_2$
Chrysoberyl, $BeAl_2O_4$
Clinohedrite, $CaZnSiO_4 \cdot H_2O$
Clinohumite, $(Mg,Fe^{2+})_9(SiO_4)_4(F,OH)_2$
Colemanite, $Ca_2B_6O_{11} \cdot 5H_2O$
Copper, Cu
Corundum, Al_2O_3
Cotunnite, $PbCl_2$
Danburite, $CaB_2(SiO_4)_2$
Diamond, C
Dolomite, $CaMg(CO_3)_2$
Elbaite, $Na(Li,Al)_3Al_6(BO_3)_3Si_6O_{18}(OH)_4$
Esperite, $PbCa_3Zn_4(SiO_4)_4$
Fluorapatite, $Ca_5(PO_4)_3F$
Fluorite, CaF_2
Franklinite, $(Zn,Mn^{2+},Fe^{2+})(Fe^{3+},Mn^{3+})_2O_4$
Galena, PbS
Gaspéite, $(Ni,Mg,Fe^{2+})CO_3$
Gold, Au
Halite, NaCl
Hardystonite, $Ca_2ZnSi_2O_7$
Haüyne, $(Na,Ca)_{4\text{-}8}Al_6Si_6(O,S)_{24}(SO_4,Cl)_{1\text{-}2}$
Hedyphane, $Pb_3Ca_2(AsO_4)_3Cl$
Heinrichite, $Ba(UO_2)_2(AsO_4)_2 \cdot 10\text{-}12H_2O$
Hematite, Fe_2O_3
Hydroxylapatite, $Ca_5(PO_4)_3(OH)$
Jadeite, $Na(Al,Fe)Si_2O_6$
Johnbaumite, $Ca_5(AsO_4)_3(OH)$
Kutnohorite, $Ca(Mn,Mg,Fe)(CO_3)_2$
Kyanite, Al_2SiO_5
Lanarkite, $Pb_2(SO_4)O$
Lazurite, $(Na,Ca)_{7\text{-}8}(Al,Si)_{12}(O,S)_{24}[(SO_4),Cl_2,(OH)_2]$
Leadhillite, $Pb_4(SO_4)(CO_3)_2(OH)_2$
Liebigite, $Ca_2(UO_2)(CO_3)_3 \cdot 10H_2O$
Magnesite, $MgCO_3$
Magnetite, $Fe^{2+}Fe^{3+}{}_2O_4$
Manganaxinite, $Ca_2Mn^{2+}Al_2BSi_4O_{15}(OH)$
Manganosite, MnO
Margarosanite, $Pb(Ca,Mn)_2Si_3O_9$
Marialite, $3NaAlSi_3O_8 \cdot NaCl$

Mattheddleite, $Pb_{20}(SiO_4)_7(SO_4)_4Cl_4$
Meionite, $3CaAl_2Si_2O_8 \cdot CaCO_3$
Meta-autunite, $Ca(UO_2)_2(PO_4)_2 \cdot 2\text{-}6H_2O$
Mimetite, $Pb_5(AsO_4)_3Cl$
Minrecordite, $CaZn(CO_3)_2$
Norbergite, $Mg_3(SiO_4)(F,OH)_2$
Norsethite, $BaMg(CO_3)_2$
Nosean, $Na_8Al_6Si_6O_{24}(SO_4) \cdot H_2O$
Nováčekite, $Mg(UO_2)_2(AsO_4)_2 \cdot 12H_2O$
Oligoclase, $(Na,Ca)Al(Al,Si)Si_2O_8$
Opal, $SiO_2 \cdot nH_2O$
Otavite, $CdCO_3$
Petalite, $LiAlSi_4O_{10}$
Phosgenite, $Pb_2(CO_3)Cl_2$
Powellite, $CaMoO_4$
Pyrite, FeS_2
Pyrolusite, MnO_2
Pyromorphite, $Pb_5(PO_4)_3Cl$
Quartz, SiO_2
Rhodochrosite, $MnCO_3$
Saléeite, $Mg(UO_2)_2(PO_4)_2 \cdot 10H_2O$
Scheelite, $CaWO_4$
Siderite, $FeCO_3$
Silver, Ag
Smithsonite, $ZnCO_3$
Sodalite, $Na_8Al_6Si_6O_{24}Cl_2$
Sphalerite, ZnS
Sphaerocobaltite, $CoCO_3$
Spinel, $MgAl_2O_4$
Spodumene, $LiAlSi_2O_6$
Strontianite, $SrCO_3$
Susannite, $Pb_4(SO_4)(CO_3)_2(OH)_2$
Svabite, $Ca_5(AsO_4)_3F$
Swedenborgite, $NaBe_4SbO_7$
Taaffeite, $Mg_3Al_8BeO_{16}$
Tilasite, $CaMg(AsO_4)F$
Topaz, $Al_2SiO_4(F,OH)_2$
Torbernite, $Cu(UO_2)_2(PO_4)_2 \cdot 8\text{-}12H_2O$
Tugtupite, $Na_4AlBeSi_4O_{12}Cl$
Turneaureite, $Ca_5[(As,P)O_4]_3Cl$
Uvarovite, $Ca_3Cr_2(SiO_4)_3$
Vanadinite, $Pb_5(VO_4)_3Cl$
Vesuvianite, $Ca_{10}Mg_2Al_4(SiO_4)_5(Si_2O_7)_2(OH)_4$
Vishnevite, $(Na,Ca,K)_6(Si,Al)_{12}O_{24}[(SO_4),(CO_3),Cl_2]_{24} \cdot nH_2O$
Willemite, Zn_2SiO_4
Witherite, $BaCO_3$
Wollastonite, $CaSiO_3$
Zeunerite, $Cu^{2+}(UO_2)_2(AsO_4)_2 \cdot 10\text{-}16H_2O$
Zircon, $ZrSiO_4$

Acknowledgements

Drafts of this chapter were submitted for comment to a number of friends and colleagues, all of whom provided valuable suggestions for improvement. I am particularly indebted to Richard Bostwick, Marilyn Grout, Isabelle Brownfield, Peter Modreski, Steven Misiur, Dan Russell, and William White for technical reviews; their comments and corrections not only strengthened the presentation but also saved me the embarrassment of immortalizing some splendid inaccuracies in print. Additional guidance was kindly given by Pete Dunn, Alix Filer, Virginia Glanzman, Elna Hauck, Erin Holmstad, Don Hoover, Don Newsome, Lawrence Rooney, David Schleicher, Yetty Verbeek, and Thomas Warren. The illustrations were drafted by Ray Sabala. Thanks to all: without your help and encouragement the preparation of this chapter would have been drudgery rather than a pleasure.

References to photographs and illustrations are printed in boldface type.